A WONDERFUL GIFT
TO THE READERS OF
"A BRIEF HISTORY OF TIME"

A WONDERFUL GIFT TO THE READERS OF "A BRIEF HISTORY OF TIME"

Bingcheng Zhao

A WONDERFUL GIFT TO THE READERS OF "A BRIEF HISTORY OF TIME"

Published in the United States of America through Amazon CreateSpace

The copyright of this book is owned by the author of this book

Copyright © 2017 by Bingcheng Zhao—the author of this book

All rights reserved, including the right of reproduction in whole or in part in any form or by any means. No part of this book may be reproduced or transmitted in any form or by any means—electronic or mechanical, photocopying, recording, without the permission from the author of this book. No part of this book may be translated into any other languages without the permission from the author of this book.

ISBN-13: 978-1544849225
ISBN-10: 1544849222

CONTENTS

FOREWORD	vii
Ch. 1: Why Difficult to Understand "Our Picture of the Universe"	1
Ch. 2: Why Difficult to Understand "Space and Time"	17
Ch. 3: Why Difficult to Understand "The Expanding Universe"	65
Ch. 4: Why Difficult to Understand the Black Holes Described by Hawking	105
Ch. 5: Why Difficult to Understand "The Origin and Fate of the Universe"	133
Ch. 6: The Certainty Mechanism behind the Uncertainty Principle	153
Ch. 7: Why Difficult to Understand "The Arrow of Time"	217
Ch. 8: Why Difficult to Understand "Wormholes and Time Travel"	227
Ch. 9: Why Difficult to Understand "The Unification of Physics"	239
Ch. 10: Not Weird if Difficult to Understand "A Brief History of Time"	255
GLOSSARY	277
ACKNOWLEDGMENTS	283
INDEX	285

FOREWORD

If you feel it's difficult to understand *A Brief History of Time*, you are not alone. Perhaps, no one could really understand that book—it turns out to be both reasonable and fair that one couldn't really understand that book.

This book (*A Wonderful Gift to the Readers of "A Brief History of Time"*) reveals the secrets of why it's very difficult to understand *A Brief History of Time*.

Most likely, what made you open the book *A Brief History of Time* was your curiosity about the big questions like: "Where did we come from? Why is the universe the way it is?" This curiosity also epitomized your interest in science. However, when you have been facing a book that couldn't be really understood at all, a bit like swallowing up a whole peanut without chewing, your curiosity and interest might be gradually fading away. As time goes on, you may even be skeptical of your curiosity and interest of this kind. This book reveals such a plain truth: it turns out that the responsibility in which you couldn't really understand *A Brief History of Time* is not on you at all, so you can reawaken or keep on this kind of curiosity and interest if you like. The great vitality of science lies with the curiosity and interest of the people from all walks of life. Such a widespread curiosity and interest is not only the inexhaustible source of the splendid and radiant charm of science, but also the great driving force of science advancement!

FOREWORD

For most people, perhaps the main purpose of reading the book *A Brief History of Time* was to look for a high level of intellectual enjoyment for their minds in spare time. Nevertheless, it seems clear enough that no one could get real enjoyment when reading a book that couldn't be truly understood at all. As days and years pass by, you may even be doubtful about whether you can find or get the satisfying enjoyment from reading popular science books. This book displays such a clear fact: you can continue to get wonderful and high-quality enjoyment from reading popular science books if you love to, because it turns out that the insurmountable difficulties in understanding *A Brief History of Time* do not arise from you!

For most middle and high school students, most probably it was the attraction of scientific knowledge and the curiosity in their minds that inspired them to read the popular science book *A Brief History of Time*. Clearly, such an attraction and curiosity relied on the support that they could understand it. But the "harsh" reality is that many of them have felt that they couldn't really understand it. This reality might have a seriously negative influence on these young people: the attraction of scientific knowledge to them is becoming weaker and weaker; their curiosity about science is being gradually blunted. This book brings them such encouraging news: it turns out that nothing is wrong on your side—please notice that the responsibility in which you couldn't truly understand *A Brief History of Time* is not on you; scientific knowledge is still a wonderful attraction, your curiosity about science is definitely correct! Your curiosity about science and your craving for scientific knowledge can bring you a bright future! And the bright future of science is not only closely related to, but also crucially depends on the curiosity and craving of young people!

Some people may want to know why I wrote this book. I like to share it with dear readers. At first, I had hesitated to write this book. However, I felt it might be better to write this book, after having known that many people had great difficulty understanding *A Brief History of Time*. Seven years ago, a smart senior engineer (he was visiting his son at my place in the summer of 2009) told me: he was very interested in Hawking's *A Brief History of Time*, so he had read its two versions several times in his native language, but still couldn't

FOREWORD

understand it. More decisively, after hearing that *A Brief History of Time* has gotten a 'nickname', the not understandable bestseller, I thought that I should write this book. Most determinedly, considering the major purpose of popular science books is to disseminate scientific knowledge, which is realized by enabling people to understand them, all my hesitation disappeared and I decided to write this book.

Personally, I greatly admire the writing style of Stephen Hawking in *A Brief History of Time:* accurate, succinct, and vivid, being an excellent example of popular science books. Frankly, not many scientists could create such a wonderful language as Hawking did in writing popular science books (a lot of people might have a similar impression as me). Yet what came along with this excellent writing were some noticeable, also inescapable, important questions about *A Brief History of Time*. Since its language was not an obstacle at all to most people, how to explain that most people couldn't really understand it? This question has been lingering in my heart. Though its language was quite easy, why was it still so difficult to understand it? This question has been hovering over my head. Since its language was easy to understand, why was its meaning not easy to follow? This question has been occupying my mind. I believe that many people might also keep asking and thinking these similar questions when they were reading *A Brief History of Time*. This book provides the clear and definite answers to the questions above.

Around the book *A Brief History of Time*, has appeared an argument or slogan like: it is still a gain no matter whether you could understand it or not; some people might believe that, primarily or probably because they found that it's really hard to understand that book. But it seems more realistic or reliable that more people would firmly believe: it is a real and bigger gain if they know about the secrets of why it's difficult to understand that book. Such a kind of authentic gain is a genuine, superb gift that this book has prepared for dear readers.

All in all, it seems safe to say that most people would agree: it couldn't be a pleasant experience when a person was going over and over, page by page, a popular science book but couldn't really understand it. On the other hand, it should be a suddenly enlightened, uniquely wonderful enjoyment—the enjoyment beyond description, when the person finally knows the root causes of why he or she

FOREWORD

couldn't really understand it. Such a kind of pleasurable enjoyment is exactly what this book brings to dear readers.

Anyway, many people may surprisingly realize that this book turns out to be a wonderful gift to them. If they feel this book should have been finished earlier, I am sorry for this book coming too late.

Bingcheng Zhao

CHAPTER 1

WHY DIFFICULT TO UNDERSTAND "OUR PICTURE OF THE UNIVERSE"

[The Window on This Chapter]

Overall, the easily perceived *reality* that it's difficult to understand *A Brief History of Time* comes from the plain *fact* explicitly pointed out by Dr. Hawking in that same book (1998 version, P. 12): general relativity and quantum mechanics cannot both be correct, because they are known to be inconsistent with each other; because these two theories essentially pervaded everywhere in *that book*.

Specifically, what is the relation between this easily perceived *reality* and this plain *fact*, regarding the picture of the universe? It seems quite reasonable and fair that one has difficulty understanding the modern picture of the universe described in *A Brief History of Time*, if general relativity, because of its crucial role in portraying this picture, turns out to be incorrect.

Does the reason leading to the inconsistency of general relativity and quantum mechanics turn out to be the root cause or "prime culprit" of *why* it's hard to understand *A Brief History of Time*? What are the direct and deepest reasons *why* these two theories are inconsistent? Let us find the direct reason; let us dig out the deepest reason!

Let us uncover the puzzle of *why* so many people have felt it's so hard to understand *A Brief History of Time*. After uncovering this puzzle, it turns out that one shouldn't have felt discouraged or disappointed for having not really understood *A Brief History of Time*.

The secrets of the inconsistency between general relativity and quantum mechanics are about to unfold right before your eyes! You are reaching the crux of this well-known inconsistency with profound implications and great mysteries. You are getting closer and closer to the key that will unlock the big puzzle of why it's difficult to understand *A Brief History of Time*.

A WONDERFUL GIFT TO THE READERS OF "A BRIEF HISTORY OF TIME"

Stephen Hawking clearly mentioned that general relativity and quantum mechanics are known to be inconsistent with each other in his book *A Brief History of Time* (1998 version, P. 12). In fact, the serious inconsistency between these two theories has been noticed for more than half a century. Moreover, this serious inconsistency owns three "terrible" names nowadays: one of the most fundamental problems in science, the deepest disaster of modern physics, and the greatest challenge in theoretical physics. For instance, this serious inconsistency has been explicitly recognized in several highly authoritative sources, including the encyclopedia of physics, *Science* Magazine (the issue of July 8, 2005, to be exact; this magazine is widely known to be one of the most authoritative, most influential and most famous publications in science), and the book *The World's 20 Greatest Unsolved Problems* (raised by more than 60 brilliant scientists—one of them is Professor Stephen Hawking, the author of *A Brief History of Time;* among them, 40 are Nobel laureates). All in all, as a well-known fact, the serious inconsistency of general relativity and quantum mechanics has become a clear *fact*, an undeniable *fact*, also an inescapable *fact!* Therefore, we have to bravely face up to such a *fact*. (Narrator: indeed, no knowledgeable experts in physics want to deny the serious inconsistency of these two theories, because they know this serious inconsistency very well.)

Overall, this serious inconsistency behaves as that general relativity and quantum mechanics cannot be put in the same package at all. When these two theories are forcibly put in the same package or intermingled together, many obviously ridiculous results, such as mathematical inconsistencies, meaningless infinite and negative probabilities, are thus created. Specifically, this serious inconsistency has two severe, also representative, clear symptoms. One is that the predictions of these two theories are always contradictory in the vicinity of black holes. Another is that the predictions from these two theories, when considering the observable universe as a whole, are not consistent at all. In addition, about and around this serious inconsistency, many other similar and candid terms have also appeared. Such terms include: 'seriously inconsistent', 'crucially incompatible', 'crucial incompatibility', 'fundamentally inconsistent', 'fundamental inconsistency', and so on. The appearance of these terms can substantially expand one's field of vision on this serious inconsistency from different angles and from different depths. This kind of expansion can considerably help one further realize the fundamentally profound consequences of this serious inconsistency.

In fact, the serious inconsistency of general relativity and quantum mechanics seems to become much more serious, even with catastrophic consequences, when viewed from a large perspective. Why? The reason is quite clear or obvious: these two theories are the main theoretical pillars of modern physics. (Narrator: as a matter of fact, no competent experts in physics deny that general relativity and quantum mechanics are the two main theoretical pillars of modern physics, because they are very familiar with these two theories.) And so, through the fundamentally and crucially important status of these two theories, one can naturally or easily think of the disastrous consequences of this serious inconsistency: what would happen if the main pillars of a large building were seriously inconsistent with each other?!

The disastrous, even perhaps devastating, consequences of this serious inconsistency seem to be warning or reminding us: we cannot afford not to dig out the reasons behind this serious inconsistency! Specifically, even merely from the practical point of view—that is, in the face of the *reality* that so many people have encountered insurmountable difficulties in understanding *A Brief History of Time*, it seems that we dare not be blind to the reasons incubating this serious inconsistency, because these two theories essentially pervaded everywhere in *that book*. Therefore, the main task of this chapter is to find and dig out these reasons (regardless of whether they are the direct reason or the deepest reason) that bring about or incubate this serious inconsistency. (Commentator: yes, the accomplishment of this task is definitely necessary! In science, it is, of course, important having noticed and recognized the serious inconsistency of two crucial theories, but it is undoubtedly far more important to unveil and realize the reasons causing this serious inconsistency. It seems no real scientists would deny that!)

Is there a quick or simple way to perceive the inconsistency of general relativity and quantum mechanics? The answer is: yes. Let us see what it is. Being a well-known or undeniable fact, general relativity is factually incompatible with energy conservation law (this law, being one of the most fundamental principles in science, tells us that energy cannot be created or destroyed, but only changed from one form into another or transferred from one object to another), which is evidenced in the *fact* that energy and matter may be created from the nothingness of the space-time of general relativity; that is, general relativity does not obey this law. On the other hand, quantum mechanics is compatible with energy conservation law, even though merely via an assumption; that is, quantum mechanics obeys this law. So these two theories follow two different rules. Consequently, it becomes obvious, even natural, or at least not a surprise at all, that general relativity and quantum mechanics are inconsistent with each other. (This is a bit like that in a sport competition, if one athlete obeyed an important rule, whereas another didn't, contradiction or inconsistency between the two athletes would appear, were the issue of violating the rules of the game set aside for a moment. This is also similar to the situation in which, if one driver observed a required rule, such as driving along the right side, but another didn't, this situation would cause contradiction or inconsistency between the two drivers.) What should be pointed out or emphasized is: clearly, even obviously, also plainly and undeniably, an indispensable requirement for these two theories to be consistent is that both of them must obey energy conservation law. This requirement, however, cannot be satisfied at all. Their inconsistency is thus unavoidable.

Then what is the direct or noticeable reason that general relativity and quantum mechanics are inconsistent with each other? Let us get it. General relativity is built on *one set* of assumptions, hypotheses and postulates (AHPs, for short); quantum mechanics is built on *another set* of AHPs (you will see these two sets of AHPs very soon). Clearly, knowing the relationship between these two sets of AHPs is the *prerequisite* for these two theories to be correctly or successfully connected with each other (that is, for them to be consistent). This *prerequisite*, however, cannot be satisfied at all, because there is no way to discover the relationship between these two sets of AHPs. In fact, there is no relationship at all between the set of AHPs underlying general relativity and the set of AHPs underlying quantum mechanics. Therefore, the direct reason of the inconsistency between these two theories is: there is no way to find and establish the relationship or connection between the two sets of AHPs that separately underlie general relativity and quantum mechanics, simply because there is no relationship at all between these two sets of AHPs. (One can easily sense or grasp this inconsistency with the following situation as an analogy. In the construction of a long tunnel across a huge mountain, the entire tunnel has been cut into several segments in order to have more working sites. Nevertheless, if each segment had its own independent rules in the parameters of elevation and direction, and if there were no way to know the relationship of these parameters among different segments, the entire tunnel could not be connected; different segments would become inconsistent with each other.)

After seeing this direct reason, some readers may feel perplexed: it is often said that both general relativity and quantum mechanics have agreed with observational results, why are these two theories still seriously inconsistent with each other? The possible perplexity of this kind seems to call for a clarification, which is provided as follows. While both theories are said to have agreed with observational results, these agreements still cannot change the very *fact* that it is definitely impossible to find out the relationship between the two sets of assumptions, hypotheses and postulates (AHPs) that separately underlie these two theories, simply because there is no relationship at all between these two sets of AHPs. That is to say, again, the *prerequisite* for these two theories to be correctly connected thus consistent with each other does not exist at all. To be clarified further: neither the observational tests of general relativity nor the experimental tests of quantum mechanics have ever involved the question of whether there is relationship or connection between these two sets of AHPs; not to mention that this question has never ever been the targeted purpose of these tests. In other words, all the observational and experimental tests of both these theories that have been done actually have nothing to do with the hard *fact* that these two theories are seriously inconsistent. (Commentator: in fact, believe it or not, it is just this direct reason that eventually unveils the long-term unanswered, also crucially important, big puzzle: why general relativity and quantum mechanics are still seriously inconsistent with each other, even though both of them have agreed with observational results.)

Now, let us see what are the two sets of assumptions, hypotheses and postulates (AHPs) that lie beneath general relativity and quantum mechanics. General relativity is based on five AHPs. (1) The postulate of 'equivalence principle' says that gravitational force has the same effect in increasing the velocity of an object as other traditional forces. This postulate is the heart and soul of general relativity. (2) The postulate of 'invariant scales of length and time' says that the scales of length and time at different points over an entire gravitational field are the same. This postulate, simply stated as 'scale invariant general relativity' by Fernando Franco, is indispensable for general relativity to deal with the issues of space and time in gravitational fields. (3) The postulate of 'photons have inertial mass' is derived from the observation that photons have momentum (a quantity of motion of a moving object), by ascribing the momentum of photons to their assumed inertial mass. This postulate is crucial for general relativity to describe the behaviors of photons (light) in gravitational fields. (4) The assumption of 'the equivalence of inertial and gravitational mass' says that the gravitational mass and inertial mass of the same object are equal to each other (the inertial mass of an object is defined based on Newton's second law. This law tells us that an object with a certain amount of mass will accelerate, or change its speed, at a rate that is proportional to the magnitude or value of the net force acting on the object; this 'a certain amount of mass' is defined as the inertial mass of the object). (5) The assumption of 'the equivalence of inertial mass and active gravitational mass' is for making gravitational mass generate the curvature of spacetime described in general relativity. Thus, one can clearly see that general relativity hires at least five AHPs indeed.

Quantum mechanics is built or based on four assumptions, hypotheses and postulates (AHPs). (1) The core assumption: all forms of energy released from electrons are in the way of discrete units or bundles called quanta, being commonly referred to as photons today (of course, what should be noticed is that one photon is released or emitted from a single electron alone). Because the core task of quantum mechanics is about how energy is released from electrons, this assumption is often referred to as the core assumption or key concept of quantum mechanics. (2) The assumption of energy conservation says that quantum mechanics obeys energy conservation law, one of the most fundamental principles in physics, also in science (this law has been mentioned above). (3) The assumption of momentum conservation tells us that quantum mechanics also obeys momentum conservation law (this law, being one of the most fundamental principles in physics, tells us that the total momentum of the objects of a system is constant if there are no external forces acting on the system. In such a system, the total momentum of two objects before a collision is equal to the total momentum of the two objects after the collision). (4) The assumption of probabilities or possibilities is about how quantum mechanics describes the observational results. With this assumption, what quantum mechanics describes or predicts is the possibility or chance (how likely) of each of the different possible results, instead of telling us a certain definite result. So, one can also clearly see that really, quantum mechanics employs at least four AHPs.

Up to here, one might have clearly realized: the trouble of the serious inconsistency between general relativity and quantum mechanics turns out to be caused by the fact that these two theories separately hire two disconnected sets of assumptions, hypotheses and postulates (AHPs). This clear realization may naturally, or easily, lead one to think of or ask the closely related question like: why does each of these two theories have to employ a set of AHPs? To this question, the answer is: the most fundamental or the deepest (also the most important) reason is that neither of these two theories is able to solve the most fundamental problem in front of itself. This answer is, of course, too short for such a crucially important question. So let us see the following analysis and discussion that lead us to this answer.

There are two solid, also tacitly admitted or undeniable, *facts* in physics. One is that general relativity is unable to resolve the most fundamental problem in front of itself, *why* space and time are variable thus relative in a gravitational field, simply because it is incapable of revealing the mechanism behind this *why*. (Narrator: since the major subject of general relativity is to tell us that space and time are variable thus relative in a gravitational field, clearly, the problem of *why* space and time are variable thus relative in such a situation is the most fundamental problem in front of this theory.) For instance, while general relativity tells us that time runs slower in a gravitational field, it is unable to tell us *why* time runs slower in such a situation.

WHY DIFFICULT TO UNDERSTAND "OUR PICTURE OF THE UNIVERSE"

In fact, this inability of general relativity has been perplexing many brilliant physicists for a long time; that is, this inability, being a tacitly admitted *fact*, is also a clearly perceived *reality*. Thus, the existence of this solid *fact* has actually become an explicitly realized or undeniable *reality*, either in truth or in essence. (Related questions and answers: wait a moment, please. Could one clearly perceive and explicitly realize, even just from the perspective of general relativity, the solid *fact* that general relativity is really unable to resolve the most fundamental problem in front of itself—*why* space and time are variable thus relative in a gravitational field? How? Answer: yes; surely and definitely! This could be easily done if he or she views or thinks over this solid *fact* conversely like the following. If general relativity had been able to resolve this most fundamental problem, its indispensable postulate of 'invariant scales of length and time', just mentioned above, would not have been necessary at all. And so, the plain *truth* that general relativity indispensably and desperately needs this very postulate can enable one to perceive this solid *fact* clearly and realize it explicitly; this plain *truth* also unavoidably shows that one has no choice but to acknowledge this solid *fact*. Of course, this solid *fact* will manifest itself even further thus become more noticeable if one thinks of or asks the simple and clear question like: if general relativity had been able to resolve this most fundamental problem, who would have looked for trouble by proposing this completely unnecessary postulate?! More than that, if general relativity had been able to resolve this most fundamental problem, four of its five assumptions, hypotheses and postulates just mentioned above, except the postulate of 'photons have inertial mass', would definitely not have been necessary in truth. Question: it is often said that general relativity has passed observational tests—such as the rotation of the long axis of Mercury's orbit, light bending around the sun, and time running slower in the gravitational field of the earth, why does such a solid *fact* still exist? Answer: because none of the existing observational tests of general relativity can change this *fact*, as in turn is because none of these tests has ever targeted to deal with this *fact* at all; not to mention that none of these tests has even intended to touch this *fact*! In other words, all these observational tests actually have nothing to do with this solid *fact*.) (Related reminder: being important general knowledge or common sense in science, observational or experimental tests themselves have neither the function nor the ability to answer the questions about the *whys* or solve the problems of the *whys*, so the task or purpose of observational or experimental tests is not to deal with these questions or these problems at all; instead, the task of answering the questions about the *whys* or solving the problems of the *whys* is, or is supposed to be, responsible by theories.

Quite obviously, such general knowledge or common sense can certainly make or help one see this solid *fact* even more clearly and explicitly.)

Another *fact* is that quantum mechanics is unable to resolve the most fundamental problem in front of itself, *why* there are quantum states, simply because it doesn't have the ability to reveal the mechanism behind this *why* at all. (Narrator: since the main task of quantum mechanics is to tell us that all forms of energy, released from electrons, are in the manner of discrete units or bundles called quanta, which are widely referred to as photons today, undoubtedly, the very problem of *why* there are quantum states or *why* there is the mechanism that enables an electron to generate photons—that is, the problem of *why* and how photons get their velocity c (the speed of light) from the electron that emits them, is definitely the most fundamental problem in front of quantum mechanics.) Specifically, quantum mechanics is unable to reveal the *quantum* mechanism of *why* and how photons, being the tiny and discrete quanta or particles of light, get their velocity c (the speed of light) from the electron that emits them. The key to sensing or realizing this inability lies with: quantum mechanics is *wave* mechanics—in fact, none of the related professional people in physics deny the *truth* that quantum mechanics is wave mechanics because they know this truth very well, whereas *wave* mechanics doesn't and can't have the function or ability to solve the problem of *why* there are *quantum* states at all. (Commentator: yes! This inability is self-evident and crystal clear. How can *wave* mechanics solve the problem of *why* there are *quantum* states?!? It is clearly impossible; it is dynamically thus essentially impossible! This is because *wave* mechanics has neither the function nor the ability to solve the problem of *why* there are *quantum* states—either in essence or in truth or in both. How can the problem of *why* there are *quantum* states be solved with *wave* mechanics?!? It is definitely impossible; it is obviously impossible! Again, this is because *wave* mechanics has neither the function nor the ability to solve the problem of *why* there are *quantum* states at all.) As a result, also as a matter of fact, this inability of quantum mechanics has been baffling many excellent physicists for a very long time in reality; that is, this inability, being a tacitly admitted *fact* in truth, has already become an explicitly perceived and realized *fact*, thus also being an actually recognized or irrefutable *fact*, or a solid *fact*.

More specifically or directly to be even more perceptible and noticeable, one could clearly perceive and explicitly realize, even only from the large perspective of quantum mechanics itself, the solid *fact* that quantum mechanics is really unable to resolve the most fundamental problem in front of itself—*why* there are quantum states, if he or she views or thinks over this solid *fact* conversely as follows. If quantum mechanics had been able to resolve this most fundamental problem, its core assumption just men-

WHY DIFFICULT TO UNDERSTAND "OUR PICTURE OF THE UNIVERSE"

tioned above, which says that all forms of energy released from electrons are in the way of discrete units or bundles called quanta (being often referred to as photons nowadays), would be utterly unnecessary; thus this core assumption would never have appeared at all. And so, the irrefutable or undeniable *actuality* that quantum mechanics indispensably and desperately needs this core assumption can definitely enable one to perceive this solid *fact* clearly and realize it explicitly; this irrefutable or undeniable *actuality* also shows that one cannot deny this solid *fact*. Not only that, if quantum mechanics had been able to resolve this most fundamental problem, three of its four assumptions, hypotheses and postulates just mentioned above, excluding the assumption of probabilities or possibilities, would not have been necessary in fact. In addition, what should be clarified or pointed out is: none of the experimental tests of quantum mechanics that have been carried out can change or affect this solid *fact*, simply because these tests have neither targeted nor intended to cope with this *fact* at all; consequently, none of these tests has the ability to change or influence this *fact* in truth. That is to say, all these experimental tests, believe it or not, really have nothing to do with this solid *fact*. (Related reminder: as just mentioned above, being important general knowledge or common sense in science, the task or purpose of experimental tests is not to answer the questions about the *whys* or solve the problems of the *whys*, because experimental tests themselves have neither the function nor the ability to do that; the task of answering these questions or solving these problems is taken on, or is supposed to be taken on, by theories instead. Quite obviously, also rather rationally, this general knowledge or common sense can substantially and noticeably help one see this solid *fact* even more clearly and explicitly, thus grasp it more tangibly and definitely.)

Up to now, we have clearly seen that the deepest reason beneath the inconsistency of general relativity and quantum mechanics turns out to be: general relativity doesn't have the ability to resolve the most fundamental problem right in front of itself, *why* space and time are variable thus relative in a gravitational field; quantum mechanics is unable to resolve the most fundamental problem directly in front of itself, *why* there are quantum states. (As a result—as a clear and noticeable result in fact, the common aspect of these two theories is that neither of them is able to resolve the most fundamental problem in front of itself, if one wants to look for their common aspect. Besides, what should be noticed is that this common aspect can rationally, even naturally, make one perceive: the fully recognized and famous inconsistency of general relativity and quantum mechanics turns out to be a clear and constant reminder of the explicit existence of the two solid facts just mentioned above. Quite obviously, such a perception is certainly a considerable help for one to face these two solid facts calmly and rationally, which is a substantial help for him or her to grasp this deepest reason firmly and effectively.)

What does this deepest reason really tell us? It reveals the two crucial, also basic, features of the inconsistency between general relativity and quantum mechanics! One is that their inconsistency turns out to be fundamentally inherent; that is, these two theories are inherently inconsistent from their roots. Another is that the crux of this inconsistency is into the essential depth of mechanism and cause, rather than on the superficial level of phenomena and effects. These two features explicitly show that the inconsistency of these two theories is not only unavoidable, but also serious, and very serious!

In fact, one can clearly see the seriousness of this inconsistency even merely through checking two available clues (both clues are fully recognized *facts* by the scientific community in physics, thus also two undeniable or inescapable *realities* in truth). First clue: there are no experimental observations that can provide any hints on how to combine general relativity and quantum mechanics. (Now, let us inspect this clue.) As pointed out above, these two theories are based on two different sets of assumptions, hypotheses and postulates (AHPs); so knowing the relationship between these two sets of AHPs is the indispensable requirement to combine these two theories correctly thus successfully. However, it is definitely not possible to find out the relationship between the set of AHPs underlying general relativity and the set of AHPs underlying quantum mechanics, simply because there is no relationship at all between these two sets of AHPs. That is, this indispensable requirement cannot be satisfied at all. Please think over: even for the same experimental observation, general relativity interprets it in this way based on one set of AHPs, whereas quantum mechanics explains it in that way based on another set of AHPs, how could it be possible to provide any hints about how to combine these two theories?! (So this clue, again which is a fully recognized *fact* or undeniable *reality*, has, in fact, sent out such a clear and serious warning to us—especially to those respected and related professional people in physics: no matter how many experimental observations have been or will be done, as long as they are unable to touch the crux or essence of the serious inconsistency between general relativity and quantum mechanics, still cannot provide any help on how to solve this serious inconsistency at all, consequently, still cannot provide any hints on how to combine these two theories.) Second clue: the yet unknown relationship between general relativity and quantum mechanics. (Let us check this clue also.) Now we can clearly see that, this unknown relationship turns out to be: there is no relationship at all between the two disconnected sets of AHPs that separately underlie these two theories. Thus, what these two clues really tell us is: in confronting this serious inconsistency, the related experts have not yet known in not only how to start with, but also in where to start from. How serious this inconsistency is! What a serious inconsistency it is!

Yet, what seems to be even more serious is that the present opinions about this serious inconsistency are seriously inconsistent too. Today there are two main camps of physicists. Camp A: use general relativity to study large-scale objects such as stars and galaxies (each galaxy has numerous stars), as well as the universe made up of almost incalculable galaxies. Camp B: use quantum mechanics to study tiny things on extremely small scales, like atoms and electrons. The representative opinion of camp A is often as: general relativity has passed a series of rigorous tests, so general relativity is fundamental; thus quantum mechanics needs to be adapted to fit general relativity. On the other hand, the dominant opinion of camp B is often like: the predictions of quantum mechanics have never been disproved after a century's worth of experiments, it thus provides a correct description of the physical world under almost all situations; therefore it is general relativity that ought to be modified to fit quantum mechanics. Now, we can see and say, the opinion of camp A is more or less to say that the fault in the serious inconsistency of general relativity and quantum mechanics is totally on the side of quantum mechanics, whereas the opinion of camp B is virtually to say that the fault is entirely on the side of general relativity. Thus, in the face of the serious inconsistency of these two theories, an interesting or ironic anecdote is that the two types of opinions about this serious inconsistency are seriously inconsistent either. (Commentator: believe it or not, the prevalence and dominance of this kind of anecdote can virtually prevent the mainstream scientists in both camps from finding out and identifying the root cause behind this serious inconsistency.) Consequently, this sort of prevalence and dominance is, actually or at least in a sense, much more serious than this serious inconsistency itself: a bit like that no rational people in the world would deny that the prerequisite of a doctor curing a patient is to find out the cause of the patient.

I cannot be sure what dear readers might think of or hear when knowing about the seriously inconsistent opinions on the serious inconsistency of general relativity and quantum mechanics. But to me, it seems that a clear voice of the intense dispute between these two theories is lingering, lingering, and lingering around my ears.

One day, general relativity and quantum mechanics began their dispute in a big conference room, in the presence of a competent arbitrator and an independent mediator. (The purpose of this dispute was to find out the exact causes of incurring the serious inconsistency between these two theories, in a hope to get possible ways of solving this serious inconsistency.)

After drinking a little water, quantum mechanics started first: "Certainly, I agree that there is serious inconsistency between general relativity and myself, since that is a fact. But I don't think the fault is on me, because my predictions have passed the experimental tests over the duration close to one century."

Followed by general relativity and it said: "Yes, of course, I also agree the reality that there is serious inconsistency between quantum mechanics and myself. But I don't think the fault is on me either, because my predictions have accurately matched observations several times over the duration about one century." After a short whisper to the arbitrator, the mediator said: "Mr. Quantum Mechanics, your statement appears to point at that the fault of this serious inconsistency seems to be on the side of general relativity, why?"

Then quantum mechanics stated firmly: "Though its predictions have matched observations several times, general relativity does not (maybe cannot) solve the most fundamental problem in front of itself, why space and time are variable thus relative in a gravitational field. This problem is obviously the most crucial and most important subject that general relativity has to deal with." "Can you give me an example?" the mediator asked. "General relativity cannot tell us why time runs slower in a gravitational field," quantum mechanics replied immediately.

The mediator then asked: "Mr. General Relativity, is that true?" "Yes, that is true. Though I have tried many times with a lot of efforts, I still couldn't figure out how to solve the problem of why space and time are variable thus relative in a gravitational field. Maybe, our human beings lack of the ability to solve this problem," general relativity answered and explained frankly. After quickly writing down something and thinking a short while, the mediator asked general relativity: "Mr. General Relativity, you have also said that the fault of this serious inconsistency is not on your side, could you please tell me why?" "Because quantum mechanics does not (or perhaps cannot) tell us why quantum states exist, also the most fundamental problem that quantum mechanics has to face," general relativity answered clearly.

The mediator then approached the arbitrator, and they whispered a few minutes. After that, the arbitrator announced: "Today's debate is over, but we have not come to a conclusion yet. We will have another debate two months later from today in the same place. In the next debate, general relativity should be prepared to answer the questions: are you able to solve the problem of why space and time are variable thus relative in a gravitational field? Is this problem the most fundamental one in your area? And quantum mechanics should be ready to answer the questions: are you able to solve the problem of why there are quantum states? Is this problem the most fundamental one in your area? See you two months later."

Two months later, the debate resumed. The mediator said to general relativity: "To be fair, Mr. General Relativity, you go first this time." Then general relativity stated honestly and solemnly: "I cannot solve the problem of why space and time are variable thus relative in a gravitational field. And this problem is the most fundamental one in my area. Maybe, we need a new theory to solve this problem."

Then the mediator turned to quantum mechanics and said: "Now it's your turn, Mr. Quantum Mechanics." Quantum mechanics answered: "I cannot solve the problem of why there are quantum states either. And this problem is the most fundamental one in my area. Probably, in order to work out this problem, a new theory is necessary."

After discussing with the mediator, the arbitrator announced: "Now, it is clear that neither of you two theories is able to solve the most fundamental problem in your own area. Based on this valid and sufficient evidence, we have clearly seen such a solid conclusion: neither of you two can be fundamentally correct, or both of you cannot be fundamentally correct. Do you agree with this conclusion?" "Yes," replied from both general relativity and quantum mechanics. The arbitrator continued: "But if you feel this conclusion is too harsh, we can say it in a mild way temporarily, as suggested by the considerate mediator: neither of you can be fundamentally understood, because neither of you can be fundamentally correct. Do you accept this mild verdict or arbitration?" "Yes, of course," both general relativity and quantum mechanics nodded their heads.

Finally, the arbitrator said: "Now we can back to the original purpose of this debate. Since neither of you two theories is able to solve the most fundamental problem in your own area, so in the serious inconsistency between you two, each side has its own fault. Please tell me yes or no." "Yes," both general relativity and quantum mechanics agreed. Soon after hearing the "Yes" from both the theories, the mediator ended the debate with: "Yes, the dispute is over now."

What cannot be further worse is the real implication of the serious inconsistency between these two theories! And this real implication can be clearly seen via the simple example as follows. When one works on a budget calculation, a straightforward common sense is: the total value of the right column must be equal to that of the bottom row, so that the same number is written at the bottom right corner. Otherwise, at least one of the two calculations is wrong—of course, not ruling out the chance that both calculations are wrong. This example thus tells us that the real implication of the serious inconsistency between general relativity and quantum mechanics is such a simple and clear fact: at least one of these two theories cannot be correct—certainly, not excluding the likelihood that both are wrong. By analogy, this real implication is also a bit like the situation in which, when a detective finds out two obviously inconsistent clues in the process of collecting evidence for investigating a murder case, at least one of these two clues cannot be valid. (Reminder: it should be noticed that this real implication, though sounding a bit "harsh", is consistent with the plain *fact* that has been explicitly mentioned by Hawking in his *A Brief History of Time:* general relativity and quantum mechanics cannot both be correct because they are known to be inconsistent with each other—1998 version, P. 12. Quite noticeably, the irrefutable existence of this plain *fact* can substantially help one perceive and realize that this real implication turns out to be not only clear and obvious, but also pretty reasonable and fair; this effect can enable him or her to face this real implication calmly and rationally.)

With this real implication, it seems that one cannot avoid the inescapable question like: which theory is incorrect, general relativity or quantum mechanics? (Let us focus on general relativity at this moment.) As pointed out above, general relativity is unable to resolve the most fundamental problem in front of itself, *why* space and time are variable thus relative in a gravitational field. For instance, general relativity cannot tell us *why* time runs slower in a gravitational field, simply because it is incapable of revealing the mechanism behind this *why* at all. One can therefore come to the solid, also unavoidable, conclusion: general relativity turns out to be incorrect. (Commentator: people shouldn't be too surprised towards this solid conclusion in reality, because it is merely an explicit confirmation of the plain *fact* that Hawking has clearly mentioned in his *A Brief History of Time:* general relativity and quantum mechanics cannot both be correct because of their inconsistency with each other. Undoubtedly, the clear existence of this plain *fact* can substantially reduce people's psychological or emotional reluctance in facing and/or accepting this solid conclusion. If one feels psychologically uncomfortable towards this conclusion, he or she can choose a mild expression for sounding less harsh, such as general relativity cannot be fundamentally correct because it is unable to solve the most fundamental problem in front of itself. However, if a theory cannot be fundamentally correct, the theory actually cannot be correct—either in essence or in truth or in both. If some people feel it's too early to accept this conclusion now, much more pieces of hard evidence will be provided in chapter three.)

With this solid conclusion, it seems not difficult that one can suddenly, but clearly, realize: it turns out both reasonable and fair that people have come across great difficulty in understanding the modern picture of the universe presented in *A Brief History of Time* (chapter one, "Our Picture of the Universe"), because general relativity is the crucial and central theory responsible for describing or portraying this picture. Such a realization is a good relief or great consolation for the people who have been perplexed or struggled for having not really understood this picture.

Not only that, with this solid conclusion as an explicit reminder, one can easily and clearly perceive: it turns out to be reasonable thus quite normal that one has had difficulty understanding *A Brief History of Time*, because its most/many parts are crucially based on general relativity. With such a perception, we have eventually revealed a secret of why it's difficult to understand *A Brief History of Time*. After revealing the secret, it turns out that people shouldn't have felt discouraged or disappointed at all for having not really understood *A Brief History of Time*.

In this chapter, we have dug out the deepest reason of why general relativity and quantum mechanics are seriously inconsistent with each other. This deepest reason is that general relativity is unable to resolve the problem of *why* space and time are variable thus relative in a gravitational field, and quantum mechanics is unable to resolve the problem of *why* there are quantum states. Having known this deepest reason, some, even many, insightful and sharp readers might want to go the extra mile. They may raise the related questions like: can we discover or develop a new theory that is able to solve the problem of *why* space and time are variable thus relative in a gravitational field? Is such a theory available at present? These questions will be concisely answered in chapter three. As to the other related question like: is there, or can we find out, a new theory that is able to solve the problem of *why* there are quantum states? It will be succinctly answered in chapter six. (The answers to these questions will enable readers to see and/or realize this deepest reason much more clearly through a sharp contrast.)

CHAPTER 2

WHY DIFFICULT TO UNDERSTAND "SPACE AND TIME"

[The Window on This Chapter]

If you have read Hawking's book *A Brief History of Time*, you might have already noticed such an explicit *fact* in it: general relativity and quantum mechanics cannot both be correct because they are known to be inconsistent with each other (1998 version, P. 12).

What does this explicit *fact* really tell us? It is quite normal that one has difficulty understanding the concept of space and time described in *A Brief History of Time*, if general relativity, due to its crucial role in forming this concept, turns out to be incorrect.

What does this explicit *fact* indicate? If general relativity turns out to be incorrect, it seems that we cannot afford not to inspect whether special relativity is correct or not, because it is the crucial basis of general relativity. (Thus the actual implication of this explicit *fact* is: no one could really understand the concept of space and time said in *A Brief History of Time*, if general relativity and special relativity, the two theories that form and determine this concept, turn out to be incorrect.)

Then how to inspect whether special relativity is correct or not? We should inspect whether it turns out to be clearly and seriously self-contradictory. (If special relativity turns out to be clearly and seriously self-contradictory, then it has to face the "harsh" reality that....)

Let us find out the answer to the big puzzle: does special relativity turn out to be clearly and seriously self-contradictory? Let us dig out the deepest reason why it's hard to understand the concept of space and time presented in *A Brief History of Time*.

Concisely and directly to the point, after reading this chapter, you will know the clear and definite answer to the big question: does special relativity turn out to be clearly and seriously self-contradictory? With this answer, you will clearly see the crux of why many people have felt it's not easy to understand the concept of space and time described in *A Brief History of Time*. After knowing this answer and grasping this crux, you can and will clearly and surely, though perhaps somewhat suddenly or unexpectedly, realize: it turned out to be rather rational or logical, thus quite understandable that one has had great, even insurmountable, difficulty understanding this concept; that is, it turned out that one shouldn't have felt discouraged or disappointed for having not really understood this concept. (This realization can easily enable one to perceive: it turned out to be reasonable and fair that a large number of people have run into difficulties in understanding *A Brief History of Time*, because *it* is crucially based on this concept. Such a perception, needless to say, is also a pleasant release or great consolation for the people who have been perplexed or struggled for having not really understood *A Brief History of Time*.)

The purpose of this chapter is to check whether or not one can really understand the concept of space and time described in Hawking's *A Brief History of Time* (chapter two, "Space and Time", 1998 version). This purpose is to be achieved through carrying out the specific task of inspecting whether special relativity turns out to be clearly and seriously self-contradictory, because it is indispensable to this concept, also the crucial foundation of general relativity, the theory that further expands this concept into gravitational fields, then into the entire universe. Thus, how to implement such a specific task, being the most important work of this chapter, is the central line running through this entire chapter.

One might ask: why do we have to take on this specific task? The answer is: there are more than enough reasons to do that. But only two major reasons will be mentioned here. First, from the perspective of the well-recognized and famous serious inconsistency of general relativity and quantum mechanics, it is necessary and/or extremely important that we should take action to inspect whether special relativity turns out to be clearly and seriously self-contradictory. Through the detailed analysis and discussion in the previous chapter, we have dug out the deepest reason or fundamental cause behind this serious inconsistency. After digging out this deepest reason, we have clearly seen, through crystal clear hard evidence, such a plain *truth* or inescapable solid conclusion: general relativity turns out to be incorrect. (Commentator: while the manifestation of this plain *truth* is a big event in science, one shouldn't feel too surprised at all, because this plain *truth* is merely a specific or frank confirmation of the explicit *fact* that general relativity and quantum mechanics cannot both be correct because of their inconsistency with each other, as clearly pointed out by Hawking in his *A Brief History of Time*—1998 version, P. 12. That is, this manifestation is totally or actually consistent with what Hawking has pointed out. As a result, such a noticeable consistency can not only considerably reduce one's surprise or psychological reluctance towards this plain truth, but also be a substantial help to our bravely facing this plain truth and its closely related profound implications.)

With this plain *truth* right in front of us, and because special relativity is the *crucial* foundation of general relativity, such an unavoidable question, which is also a closely related profound implication of this plain truth, will thereby appear: is special relativity a correct theory? And more explicitly thus more definitely, does special relativity turn out to be clearly and seriously self-contradictory, considering this theory has agreed with observational results? (Narrator: in science, a correct theory must at least satisfy two indispensable and crucial requirements *simultaneously*—it must agree with observational results; and it must not be clearly and seriously self-contradictory. Otherwise, if a theory can meet *only* one of these two requirements, the theory cannot be correct—to be rational, objective and fair, being one of the self-evident and undeniable basic standards in science. Moreover, such a basic standard is so simple and clear, so obvious and plain that it has become a well-known and irrefutable rule in the field of science; thus it seems quite safe if one concludes that all real scientists would have no objection to this irrefutable rule. In fact, it seems to be a rather rational and realistic estimate that all knowledgeable or qualified scientists, including physicists of course, would totally agree with and actively uphold this irrefutable rule, either in their minds or in their actions.)

Second, from the large perspective of why so many people have felt it's so hard to understand *A Brief History of Time*, it seems wise that we should have a close scrutiny of whether special relativity turns out to be clearly and seriously self-contradictory. In the face of the *reality* that so many people have encountered insurmountable difficulties in understanding *A Brief History of Time*, it seems neither irrational nor inappropriate if one asks the question like: do these undefeated difficulties rest with the impassable obstacles coming from that people couldn't really understand the concept of space and time said in *it*, because this concept is the crucial basis of *that book*? (That is, it would be impossible that one could really understand *A Brief History of Time*, if he or she couldn't truly understand the concept of space and time described in *it*.) With this question kept in mind, it thus becomes clear that we ought to take action to scrutinize whether special relativity turns out to be clearly and seriously self-contradictory (because special relativity is the crucial foundation of general relativity; because special relativity and general relativity are the two theories that determine the concept of space and time described in *A Brief History of Time*).

Through these two reasons, one can clearly see: it seems that we cannot afford not to take on the task of inspecting whether or not special relativity turns out to be clearly and seriously self-contradictory. With these two reasons as a noticeable reminder, it becomes clear that there is no reason for not carrying out this specific task, because it points to the purpose of this chapter. (Commentator: more directly, in the face of the explicit *fact* that general relativity and quantum mechanics cannot both be correct because of their mutual inconsistency; and in front of the undeniable *reality* that special relativity is the *crucial* basis of general relativity, it seems clear enough that no rational people would/should have objection to the action to inspect whether special relativity is correct or not. And more specifically, after taking account of the factor that special relativity is believed to have agreed with observational results, it thereby becomes sure enough that no rational people would/should think it is not necessary to check on whether special relativity is clearly and seriously self-contradictory. This is a bit like that, if theory A is the crucial foundation of theory B that is clearly and seriously inconsistent or contradictory with theory C, then no rational people would think it's irrational to examine whether theory A is self-contradictory or not.) (Of course, it is the eventual result after finishing this task that will have a final say on whether the action of carrying out this task is really worthy or not, truly wise or not. You will see this eventual result very soon.) So, let us start our work on this specific task!

Please don't be scared away by this specific task: it is quite straightforward from this specific task to the purpose of this chapter. In order to check whether one could really understand the concept of space and time described in *A Brief History of Time* (again, chapter two, "Space and Time"), we need to examine whether we could truly understand special relativity and general relativity, the two theories that determine this concept. In order to check whether or not one could really understand general relativity, we need to examine whether we could truly understand special relativity, the crucial foundation of general relativity. In order to check whether or not one could really understand special relativity, we need to inspect whether special relativity turns out to be self-contradictory or not. Therefore, the central idea of this chapter is: if special relativity turns out to be clearly and seriously self-contradictory, then no one could really understand it; thus it is actually impossible to imagine that one could truly understand general relativity (because it is crucially based on special relativity); it is thereby reasonable and fair that one has encountered many challenges in understanding the concept of space and time presented in *A Brief History of Time*. (Narrator: the simple logical reasoning above, which is also the concise outline of this chapter, further indicates, even underlines, the great importance and/or noticeable necessity of knowing whether special relativity is clearly and seriously self-contradictory or not.)

How to inspect whether special relativity turns out to be self-contradictory or not? Of course, we need to examine whether its *core* concepts are self-contradictory or not. What are its core concepts then? Answer: the core concepts of special relativity are length contraction and time dilation, respectively for interpreting length becomes shorter and time runs slower that appear in the situation of high speed. In many materials on special relativity, length contraction and time dilation are very often said to be the most crucial core or innermost core of special relativity; and some of them go even further by saying that length contraction and time dilation are more or less equivalent to special relativity. Therefore, in order to inspect whether special relativity turns out to be self-contradictory or not, what we need to do is to examine whether length contraction and time dilation turn out to be self-contradictory. If length contraction and time dilation turn out to be factually incompatible thus essentially contradictory with each other, then they are clearly and seriously self-contradictory; that is, special relativity turns out to be clearly and seriously self-contradictory. (Narrator: undoubtedly, if special relativity turns out to be clearly and seriously self-contradictory indeed, then no one could really understand it. Accordingly, no one could truly understand the concept of space and time based on it.)

Now let us examine whether length contraction and time dilation are factually compatible or not. At present, in describing length contraction and time dilation, they are factually separate from each other! For instance, in demonstrating length contraction, the following scheme is often involved (Fig. 2.1, next page). An object, which can be a spaceship (say, its rest length is 250 feet and its height is 50 feet), approaches a 50-foot vertical line when its velocity is very close to the speed of light, because the original 250-foot rest length is shortened close to zero by length contraction. (If there is a meter stick in *the* spaceship, this meter stick is shortened close to zero either by length contraction, because it must be placed along the length direction of *the* spaceship for measuring its length.) On the other hand, in illustrating time dilation, three identical clocks are used: two are at rest in a stationary spaceship, and one is in a moving spaceship (Fig. 2.2, next page). It should be well aware that the clock in the moving spaceship is *always* identical as the two at rest, even when the velocity of the spaceship is increased up to close to the speed of light (this spaceship is thus forced to throw away its length contraction utterly: it does not experience length contraction at all). (One can easily get an animated description of length contraction and time dilation like the above, from today's web with the key words such as 'length contraction' and 'time dilation' etc.) Please closely notice what I want to emphasize and point out: the clock in the spaceship moving at high speed does not experience length contraction at all! And moreover, in the various textbooks on special relativity, the clock moving at high speed *never* experiences length contraction either (so, many professionals in physics must or might have already noticed and perceived this obvious feature!). Here a crucially important question is directly coming towards us: can a meter stick, which is used to measure length or space, be put together with a clock in the *same* spaceship moving at high speed? If not, why? The answer will be seen in the coming two paragraphs.

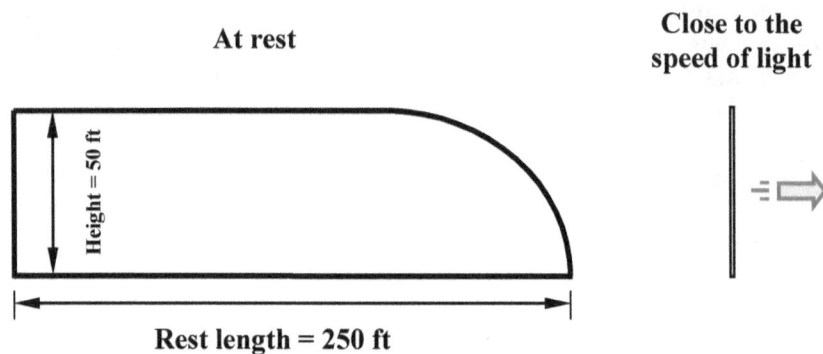

Figure 2.1, the length contraction in the theory of special relativity, which is responsible for interpreting the phenomenon of length becoming shorter that appears at high speed.

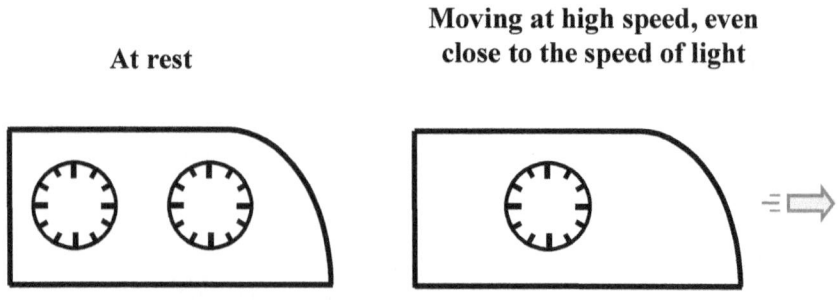

Figure 2.2, the time dilation in the theory of special relativity, which is responsible for interpreting the phenomenon of time running slower that appears at high speed.

WHY DIFFICULT TO UNDERSTAND "SPACE AND TIME"

What would happen if the description of length contraction and time dilation were presented in the manner that a meter stick and a clock are put in the *same* spaceship moving at high speed? One of the following three obviously and seriously contradictory situations would definitely appear within the paradigm or stereotype of special relativity (Fig. 2.3, next page)! Situation 1: according to the description above, the meter stick would experience length contraction (in the same way as the spaceship illustrated in Fig. 2.1, because the meter stick has to be placed along the length direction of the spaceship in order to measure its length); whereas the clock, which is certainly made up of a certain amount of mass—just like the meter stick, would not experience length contraction at all. This is clearly self-contradictory. Situation 2: admitting the clock would experience length contraction (for being consistent with the meter stick), then this clock would be squeezed into "0" even "l" shape from original "O" shape by its length contraction. This is clearly and directly contradictory with what special relativity has said about time running slower via its time dilation— the clock moving at high speed never experiences length contraction at all in special relativity (*even if temporarily putting aside the issue of whether such a seriously squeezed clock could still work or not. Of course, it seems rather rational if one has the reasoning: most likely, such a seriously squeezed clock couldn't work, because it would be very skeptical that it could still be able to measure time accurately. That is, such reasoning seems to be reasonably acceptable to most people, except those who want to refuse it on purpose with some extreme excuses*). Situation 3: denying the clock would experience length contraction (for being consistent with the requirement of special relativity), then the meter stick could/would not experience length contraction either. This is clearly incompatible or plainly contradictory with another requirement of special relativity: a meter stick must experience length contraction at high speed.

A WONDERFUL GIFT TO THE READERS OF "A BRIEF HISTORY OF TIME"

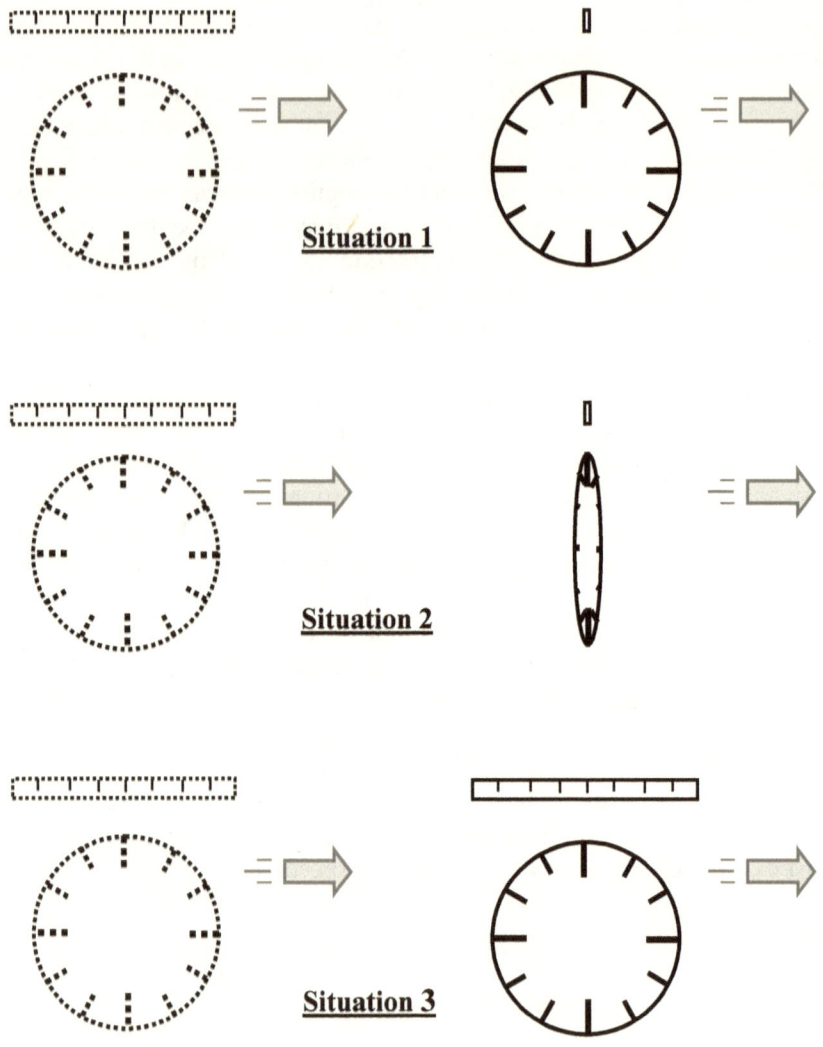

Figure 2.3, a meter stick and a clock cannot be put together in the same spaceship moving at high speed, according to the theory of special relativity.

So each of these three situations shows that special relativity has caused obviously ridiculous results! In fact, only these three situations could occur within the paradigm of special relativity. Moreover, what ought to be pointed out and noticed is: meter stick and clock are *never* put together either in the various textbooks on special relativity when its length contraction and time dilation are interpreted; that is, many insightful professional people in physics, when teaching and studying the theory of special relativity, must or might have already realized such a noticeable feature. We have therefore seen a crucially important *reality*—a clear *reality*, an unavoidable *reality*, also an undeniable *reality*, which is that within the paradigm of special relativity, a meter stick and a clock cannot be put together in the *same* spaceship moving at high speed. (The focus or crux of this *reality* is: according to the theory of special relativity, a meter stick must experience length contraction in the state of high speed, whereas a clock must NOT experience length contraction in such a state.)

Then what does such a *reality* really tell us? Or what are the actual implications of such a *reality*? Clearly and obviously, length contraction and time dilation (respectively for interpreting that length becomes shorter and time runs slower, the two phenomena that appear in the situation of high speed, as just mentioned above) are factually incompatible thus essentially contradictory, because they would directly deny each other if met together. Evidently and undeniably, length contraction and time dilation, therefore, turn out to be clearly and seriously self-contradictory. (Narrator: stated plainly, this clear and serious self-contradiction is explicitly reflected in the unavoidable, unconcealed *fact* that length contraction and time dilation don't and can't coexist at all! This is because: in order to demonstrate length contraction with a spaceship moving at high speed, a clock, which is used to measure time thus to show time dilation, is utterly not allowed to be put in *the* very spaceship; in order to illustrate time dilation, the length contraction of a spaceship moving at high speed is forcibly given up!) Inescapably and inevitably, special relativity turns out to be a theory that is clearly and seriously self-contradictory, simply because its two *core or key* concepts, length contraction and time dilation, turn out to be clearly and seriously self-contradictory. (Commentator: in science, any theory, including special relativity of course, as long as its two core or key concepts turn out to be clearly and seriously self-contradictory, the theory is clearly and seriously self-contradictory, being a plain rationale or self-evident truth, thus also being a clear and irrefutable rule. In fact, this rule is so obvious that it is quite noticeable to most people; that is, it is both rational and realistic if one concludes that this rule is reasonably acceptable to most people, even if not everyone. In other words, it is not only obviously rational but also definitely sensible to believe that no real scientists would deny such an irrefutable rule. Moreover, if anybody disagreed with this irrefutable rule in the name of a scientist, he or she would become a laughingstock of all rational people; in reality, it seems clear enough that no real scientists want to challenge this irrefutable rule.)

So, succinctly and accurately, the *reality* revealed above (which is that special relativity utterly does not allow a meter stick and a clock to be put together in the *same* spaceship moving at high speed) explicitly shows the three inherently connected facts as follows. Length contraction and time dilation are factually incompatible thus essentially contradictory (a basic fact) → they are clearly and seriously self-contradictory (a solid fact) → special relativity turns out to be clearly and seriously self-contradictory (a clear fact).

Related questions and answers: what is the relation of these three facts? Answer: from the basic fact to the solid fact to the clear fact is inescapable or inevitable, thus the only route. In other words, once one has understood, through the hard evidence displayed above, the basic fact, he or she will have no difficulty grasping the solid fact and the clear fact. Question: what are the actual or inescapable implications of this clear fact? Or more directly thus more explicitly, what is the relation between this clear fact and the purpose of this chapter? Answer: in science, there is a self-evident truth or fully recognized common sense, thus also an obvious and undeniable rule, which is that no one could really understand a theory that turns out to be clearly and seriously self-contradictory; it seems quite safe to believe that all real scientists are familiar with this rule very well, or totally agree with it. Thus, with the clear fact that special relativity turns out to be a theory that is clearly and seriously self-contradictory, it becomes clear enough or rather rational, at least not irrational at all, that one couldn't really understand special relativity in truth (that is, it turns out to be quite reasonable and fair if one has felt that he or she couldn't really and truly understand special relativity, either in fact or in essence). Accordingly, the concept of space and time described in *A Brief History of Time* couldn't be really understood, simply because this concept is crucially based on special relativity. So this clear fact directly points to the purpose of this chapter; this purpose is about checking whether or not one can really understand the concept of space and time described in *A Brief History of Time* (chapter two, "Space and Time", 1998 version), as stated at the beginning of this chapter. And moreover, the key to clearly understanding this purpose lies in explicitly realizing these three facts. Question: then how could one realize these three facts even more clearly and definitely from other noticeable angles, besides the irrefutable or hard evidence presented above? Answer: the pieces of specific information in the coming several paragraphs can provide a substantial help from several other noticeable angles.

WHY DIFFICULT TO UNDERSTAND "SPACE AND TIME"

Overall and most of all, one can more clearly witness the basic fact that length contraction and time dilation are factually incompatible thus essentially contradictory with each other, through the intense dispute between a spaceship about to travel at high speed and a clock inside it. (The spaceship represents length contraction, and the clock stands for time dilation in their dispute.) Before leaving, the spaceship started first: I would experience length contraction according to special relativity, and would approach a vertical line as my original height when my velocity is close to the speed of light. The clock refuted seriously: no, no, in that situation I would be squeezed into "0" even "l" shape by your length contraction from the original "O" shape; this would utterly put me in a position that is directly contradictory with the requirement of special relativity on me—according to special relativity, I must NOT experience length contraction at high speed. Besides, I am afraid this would make me unable to measure time. So let me say first: I want to work properly and measure time accurately, according to the requirement of special relativity on me—that is, I will NOT experience length contraction. The spaceship argued without delay: no, no, in that situation I would not experience length contraction at all, as required on me by special relativity. Being upset and angry, the spaceship yelled to the clock: get off me; if not, you will destroy my length contraction! Being badly annoyed and hurt, the clock cried: of course, I will get off, because otherwise your length contraction will make me unable to measure time, and ... and I will become useless, and ... and worthless! (Narrator: quite obviously, through the intense dispute above, this basic fact manifests itself even further, thus becomes much more noticeable; this effect is, of course, substantially helpful for one to realize this basic fact even more clearly and definitely. This realization, in turn, can enable one to be well aware of the solid fact that length contraction and time dilation turn out to be clearly and seriously self-contradictory. With this solid fact kept in mind, it seems clear enough or rather rational that one can explicitly understand the clear fact that special relativity turns out to be clearly and seriously self-contradictory indeed, because length contraction and time dilation are its two core or key concepts, as pointed out above.)

The passionate dispute above also reminds me of the story about the idiom of self-contradiction, which is the fable about spear and shield, and I like to share it with dear readers here. (You know, in ancient times, spear and shield were two different weapons used in battle. Spear was used to pierce enemy, whereas shield was for protecting one from being damaged by a sharp weapon like a spear.) More than two thousand four hundred (2400) years ago, a man in the State of Chu (1030—223 BC, in today's central and southern China) was selling both spears and shields on a market. He was loudly praising his shields one day. "My shield is so strong that nothing can pierce it through." After a short while, he was loudly bragging about his spears also. "My spear is so sharp that it can pierce through anything." "Then what will be happening, if your spear pierces into your shield?" he was asked very quickly. The man didn't know how to reply, embarrassed and left silently (because his shield and spear couldn't coexist at all—a note from the author of this book). (Commentator: after reading this fable, some, even many, insightful and sharp readers might have perceived that this fable is actually a considerable help for one to chew over and digest the solid fact that length contraction and time dilation turn out to be clearly and seriously self-contradictory, because they don't and can't coexist either. Such a perception is definitely correct! Not only that, this perception can substantially help one in distinctly and explicitly realizing the clear fact that really, special relativity turns out to be clearly and seriously self-contradictory, as the obvious and inescapable consequence of its two core or key concepts, length contraction and time dilation, turn out to be clearly and seriously self-contradictory.)

On seeing the solid *fact* that length contraction and time dilation (respectively for interpreting length becomes shorter and time runs slower at high speed) turn out to be clearly and seriously self-contradictory, some dear readers may feel a surprise at first; other people, especially some of those respected professionals in physics, might not be psychologically comfortable or accustomed in the beginning. So a clarification is provided here. While separately and individually, and viewing length contraction and time dilation one by one, it looks as if they are compatible or consistent, just because their incompatibility or inconsistency doesn't emerge in such a situation. However, when length contraction and time dilation are pieced together, and considered at the same time, really they are factually incompatible thus directly contradictory with each other—that is, they are clearly and seriously self-contradictory indeed. (Related questions and answers: what bitter lesson can we draw or learn from such a solid *fact?* Answer: comprehensive and careful inspection of the different aspects of a theory is fundamentally important in science, if the theory deals with two or more different aspects; and I believe many readers may have better answers than me. Question: what is the crux that length contraction and time dilation turn out to be clearly and seriously self-contradictory then? Answer: length contraction cannot tell *why* length becomes shorter, and time dilation cannot tell *why* time runs slower.) (Having seen this answer, it seems quite reasonable or at least not irrational at all that some insightful and sharp readers may be wondering: can we find a new theory that is able to uncover the secret mask of *why* length becomes shorter and *why* time runs slower? If so, one can see more clearly, through a sharp contrast, the solid *fact* that length contraction and time dilation turn out to be clearly and seriously self-contradictory indeed. Dear readers will see such a new theory in this chapter after a short while, but please don't hurry at this moment; I hate to divert your attention away from the central line of this chapter at this point.)

In addition to what we have discussed above, there is a clear signal that can easily direct us to go to the clear *fact* that special relativity turns out to be a theory that is clearly and seriously self-contradictory. Such a signal turns out to be the well-known *fact* that general relativity and quantum mechanics are seriously inconsistent or contradictory with each other (for the related or detailed information about this well-known fact, please go back to chapter one if necessary). Why? This well-known fact can naturally or easily enable one to perceive and realize: since special relativity is the *crucial* foundation of general relativity, and since general relativity is seriously inconsistent or contradictory with quantum mechanics, the existence of this clear fact is or becomes quite normal; that is, one shouldn't feel too surprised towards this clear fact at all. (This is quite similar to: if theory A is the crucial basis of theory B that is clearly and seriously inconsistent or contradictory with theory C, then it shouldn't be a surprise that theory A turns out to be clearly and seriously self-contradictory.) Moreover, this signal is actually a very bright and eye-catching indicator, because this well-known fact has already become very important general knowledge or common sense in physics; and so, it's not only quite rational but also pretty safe to conclude or believe that all the knowledgeable professional people in physics are familiar with this well-known fact very well. Thus, if some people, especially some of the respected professionals in physics, still feel a big surprise on seeing this clear fact, please think of and think about this well-known fact, which can considerably lessen or buffer their possible surprise; such an effect can greatly help them in bravely facing this clear fact. More candidly or directly, if some people are still psychologically reluctant or resistant towards this clear fact, they can think of and think over this well-known fact, which, like a dose of good medicine, can effectively remove or substantially reduce such a kind of reluctance or resistance. (Commentator: since this clear fact has already been shown by the irrefutable or hard evidence presented above, and since this evidence is easy to understand; along with that this well-known fact turns out to be a clear signal that can easily guide us to go to this clear fact, it seems quite reasonable and fair to believe that one could be well aware of this clear fact up to now.)

Having seen the clear *fact* that special relativity turns out to be clearly and seriously self-contradictory, it seems natural or unavoidable if one asks the question like: what is the real implication or actual consequences of this clear fact? Or more specifically thus more directly, with this clear fact, can special relativity be a correct theory? The answer to this kind of question is crystal clear: special relativity turns out unable to be correct, because this answer is sufficiently and conclusively determined by (or comes from) the following irrefutable evidence. This evidence is such a clear, even obvious, also self-evident and undeniable, rule in science: any theory, as long as it is clearly and seriously self-contradictory, cannot be correct in truth. Moreover, this rule has been universally accepted as one of the most important basic criteria or standards in science, because of the following two plain reasons. No real scientists should come to the conclusion that claims a theory that turns out to be clearly and seriously self-contradictory is a correct theory; no rational people in the world could imagine, would believe, or admit a theory that turns out to be clearly and seriously self-contradictory can be a correct theory (as a matter of fact, it is neither correct nor true, neither rational nor valid if anyone asserts a theory, which is clearly and seriously self-contradictory, as a correct theory, simply because such an assertion is not, and cannot be, the fact at all). Such a self-evident and universal rule is, of course, totally applicable to the theory of special relativity; and this application inevitably results in the answer above. In addition, since this rule is self-evident and universal, this answer is inevitable or unavoidable; because this rule is clear, even obvious, this answer is easy to understand. All in all, and concisely, by such a plain and indisputable rule, this answer is simply a natural or obvious outcome of this clear fact. In other words, once one has grasped this clear fact, he or she will have no difficulty perceiving and realizing this answer.

If this answer is still a surprise to some people (for example, some of the respected professional people in physics may feel this answer comes too suddenly), they can view it from two other noticeable angles. (The specific information from these angles can greatly help them in effectively reducing or substantially buffering their possible surprise, which is very helpful for them to face this answer calmly and rationally.) First angle, from what Stephen Hawking has said that general relativity and quantum mechanics cannot both be correct, because these two theories are known to be inconsistent with each other (*A Brief History of Time*, 1998 version, P. 12), one could logically come to the solid conclusion that special relativity cannot be correct. Why? How? By comparison, if two theories are inconsistent with each other, they cannot both be correct; then if the two *core* concepts of the same theory turn out to be clearly and seriously inconsistent or contradictory with each other, naturally this theory cannot be correct. So by this reason, it seems clear that people shouldn't feel surprised towards the answer that special relativity turns out unable to be correct. Second angle, when viewed from the perspective of the combination of the two known elements (one is that general relativity turns out to be incorrect, as mentioned earlier in this chapter; another is that special relativity is the crucial foundation of general relativity), this answer also appears rational and fair—at least to a great extent. That is, by this reason, it becomes clear that one shouldn't feel surprised towards this answer either. Moreover, when these two reasons are seen together, it seems there is no reason that this answer is still a surprise. In fact, the existence of these two reasons can enable one to realize that this answer turns out to be quite reasonable and fair.

So up to now, we have gone along the following route. Length contraction and time dilation are factually incompatible thus essentially contradictory (a basic fact) → they are clearly and seriously self-contradictory (a solid fact) → special relativity turns out to be clearly and seriously self-contradictory (a clear fact) → it turns out that special relativity cannot be correct (an unavoidable fact or natural outcome). (Related question and answer: what is one of the best ways to understand these four facts easily and effectively? Answer: as long as one has clearly comprehended the basic fact, he or she can easily and quickly realize the other three facts in a row, simply because they are utterly the inescapable, also inevitable, implications of this basis fact.)

In facing the unavoidable fact that special relativity turns out to be an incorrect theory, one might be bewildered: it is said that special relativity has passed observational tests, why is it still incorrect? About the possible bewilderment of this sort, please allow me to clarify a little bit. This clarification focuses on: all observational tests of special relativity still cannot change the solid *fact* that its two core concepts, length contraction and time dilation, turn out to be clearly and seriously self-contradictory (this is a bit like that all observational and experimental tests of general relativity and quantum mechanics still cannot change the well-known *fact* that these two theories are seriously inconsistent or contradictory with each other). In fact, all observational tests of special relativity have never ever targeted to deal with this solid fact; not to mention that none of these tests has even intended to touch this solid fact at all! That is, all these observational tests actually have nothing to do with this solid fact, believe it or not. Therefore, it is the objective and real existence of this solid fact that determines the unavoidable fact that special relativity turns out unable to be correct. (Related question and answer: why and how does this solid fact determine such an unavoidable fact? Answer: in science, a correct theory must at least satisfy two crucial and indispensable requirements *at the same time*— it must agree with observational results; and it must not be clearly and seriously self-contradictory, as pointed out earlier in this chapter. Since special relativity turns out to be a theory that can meet *only* one of these two requirements, it cannot be correct.) (Narrator: if some people, especially some of the respected professionals in physics—particularly some of those experts on special relativity, are not psychologically comfortable or accustomed towards the newly revealed *fact* that special relativity turns out unable to be correct, probably because this fact comes too suddenly in their eyes, the friendly information in the coming paragraph can provide a considerate consolation for them, at least for the time being or to some extent.)

Even though the clear fact, which is that special relativity turns out to be clearly and seriously self-contradictory, has no choice but to tell us the unavoidable *fact* that special relativity cannot be correct, it seems not unrealistic that not all people are psychologically ready or happy to face this unavoidable fact—some esteemed professional people might not be psychologically comfortable or accustomed towards this unavoidable fact at first. This kind of possible reaction seems to be understandable or sympathetic—at least in a sense, considering it cannot be a pleasant experience for some people to face the fact that their cherished theory turns out unable to be correct. For that reason, and only for that reason, at present I still advocate that it is enough to realize such a plain fact: special relativity is actually a theory that couldn't be really understood, simply because it turns out to be clearly and seriously self-contradictory. (Commentator: yes; because this plain fact is clearly and sufficiently determined by such a self-evident rationale that is a fully recognized basic principle in science: no one could really understand a theory that turns out to be clearly and seriously self-contradictory. In fact, this self-evident rationale is so obvious that it has become an irrefutable or incontrovertible rule in the field of science. And so, if someone, even in the name of a scientist, claimed that he or she could really understand a theory that turns out to be clearly and seriously self-contradictory, he/she would definitely become a big laughingstock of all rational people.) Fortunately, such a realization is, in effect, sufficient to grasp the central idea of this chapter, because this plain fact is exactly the core point of this central idea. (Speaking of the central idea of this chapter, please allow me to remind what has appeared earlier. This central idea is: if special relativity turns out to be clearly and seriously self-contradictory, then no one could really understand it; thus it is actually not possible to imagine that one could truly understand general relativity, because it is crucially based on special relativity; therefore, it turns out to be quite reasonable and fair that one has encountered many challenges in understanding the concept of space and time said in *A Brief History of Time*, because this concept is determined by special relativity and general relativity.)

Narrator: up to here, and at this moment, dear readers may have two typical reactions towards the uncovered fact that special relativity turns out to be a theory that could not be really understood (because this theory turns out to be clearly and seriously self-contradictory, which is determined by the revealed or shown fact that its two core or key concepts are clearly and seriously self-contradictory). Some readers may want to know more about special relativity actually being a theory that could not be truly understood; others might still feel a big surprise (probably because this uncovered fact is too new or too radical in their eyes, along with it comes too suddenly in their minds). Either of these possible reactions seems to need a noticeable window on special relativity, so that more people can further see whether special relativity could really be understood or not. Such a kind of window is the famous Twin Paradox, which is a highly influential story not only in special relativity but also in science fiction. So let us together inspect whether the Twin Paradox is truly a paradox or not in the next several paragraphs.

Another important thing to do in checking on whether special relativity is a theory that could really be understood or not is to inspect whether the famous Twin Paradox is truly a paradox. (When a concept, an idea, or a story is self-contradictory, we often refer to it as a paradox.) If this Paradox turns out to be truly a paradox, the side on which special relativity could not be really understood will get an important demonstration or testimony, because this Paradox comes from special relativity. On the contrary, if this Paradox is not a paradox, the side on which special relativity could be understood will still keep an important reason. Thus the basis or rule of judging this check is objective, rational, clear and fair to both sides.

Well, as a preparation, let us review the famous Twin Paradox at the start. The Twin Paradox arises from the effect of time dilation (which is the effect of time running slower in the situation of high speed, because in special relativity, time dilation is responsible for interpreting the phenomenon of time running slower in such a situation, as pointed out earlier). It says there are twin brothers. One brother stays at home in a place of a slow speed (on the earth, for example). The other brother goes away for a long trip in a very fast spaceship. The "slow" twin ages considerably: has lost most of his hair. The younger twin returns still young: still has all of his hair. This thought situation is called the famous Twin Paradox. (The story of the Twin Paradox has appeared in many materials on talking about or explaining the theory of special relativity. A lot of people might have already known this prevalent and famous story from various sources, I believe.)

In order to check on whether the famous Twin Paradox is truly a paradox or not, another necessary preparation is to know about the concept or idea of relativistic mass (that is, mass increase with speed), an important concept formed within the paradigm of special relativity. This concept says that: the mass of an object increases with the increase in its velocity, and the mass of an object becomes infinite large when the object infinitely approaches the speed of light. So 'relativistic mass' is often simply said as 'rest mass is least'. The name of relativistic mass has frequently appeared in the various materials on special relativity. (As a reminder or comparison, Hawking also mentioned this concept in his book *A Brief History of Time* as that: "As an object approaches the speed of light, its mass rises ever more quickly....")

Now we are ready to check whether the famous Twin Paradox is truly a paradox or not. How to do that? It will be checked together with relativistic mass (mass increase with speed) that comes from special relativity, just like the Twin Paradox. Let us say, each of the twin brothers has the rest mass of 75 kilograms (165 pounds), and the spaceship, in which one brother of the twins would stay, would be traveling at the velocity of 98 percent of the speed of light. Now let us see what would happen to the brother traveling away.

According to relativistic mass, the mass of the brother in the spaceship would increase from 75 to 377 kilograms (165 to 830 pounds)! How could this dear brother still be alive after experiencing such an explosive increase of mass?!? Clearly, no rational people can think this result is reasonable. When he returned from this trip—if he could still be alive, his mass would shrink from 377 to 75 kilograms (830 to 165 pounds)! How could this dear brother still be alive after experiencing such a rapid decrease of mass like a quick evaporation?!? Also clearly, no people with clear thinking can imagine this result is valid. Therefore, the Twin Paradox, when considered together with relativistic mass (mass increase with speed), turns out to be a self-contradictory story. That is, the famous Twin Paradox turns out to be truly a paradox as long as it is considered simultaneously with relativistic mass; and no other excuses or "explanations" can change such a clear and definite fate. (Related question and answer: your analysis and conclusion above might cause the argument like: no, no, the Twin Paradox is just for explaining the effect of time dilation, so you shouldn't put this Paradox together with relativistic mass. What is your response to that sort of possible arguments? Answer: the very *fact*, which is that the Twin Paradox and relativistic mass couldn't be put together, actually and exactly shows that special relativity turns out to be a self-contradictory theory, thus couldn't be really understood. Please notice that both the Twin Paradox and relativistic mass are the products of special relativity!)

In fact, when considering together with length contraction (which is responsible for interpreting the phenomenon of length becoming shorter that appears at high speed), one can see even more clearly that the famous Twin Paradox turns out to be really and truly a paradox. For the brother in the spaceship traveling at the velocity of 98 percent of the speed of light, his mass would increase from 75 to 377 kilograms, thus would be 5.03 times his original mass (this would enlarge his area from 1 to 5.03), but his length would only be 1/5.03 of his original length (this factor would also expand his area by the ratio of 1 to 5.03). So the area of this brother would be about 25 times his original size; this dear, but poor, brother would thereby be squeezed into a big but very thin board, like a large piece of plywood.

Now, one can clearly see that the Twin Paradox, when considered together with relativistic mass (mass increase with speed) and length contraction, turns out to be a truly self-contradictory story indeed. Moreover, neither excuses/pretexts nor "explanations" have the magic power to change this definite fate! Consequently, also unavoidably and undeniably, the very *fact*, which is that the Twin Paradox, relativistic mass and length contraction cannot be put together at all—they three are utterly not compatible, is a clear reflection or concrete embodiment that special relativity turns out to be a theory that is clearly and truly self-contradictory, thus could not be really understood. (Commentator: now, it becomes clear that the prevalent opinion, which says that the Twin Paradox seems not paradox as long as one abandons the idea of absolute time, turns out to be simply the consequence of merely staring at the effect of time dilation—the effect of time running slower at high speed, while utterly forgetting relativistic mass and length contraction. Of course, such an opinion turns out to be neither valid nor correct in fact; please remember that the Twin Paradox, relativistic mass and length contraction are all the three products of special relativity.)

Well, people will see the new story of the famous Twin Paradox is quite different from the original one. I think we can rewrite this new story using the results just revealed above. Let us say, in the United States there was a senior professor in physics, also an expert on special relativity, whose name was George. George had two twin sons, John and Mark. They almost finished their study for master degree, and planned to continue their education for a Ph.D. degree. On a Friday evening, they three were chatting around a table after dinner.

George started first: "Since the major of both of you is physics, I am sure you guys must be familiar with the famous story of the Twin Paradox."

"Yes, we are," John and Mark answered quickly.

"Then let's talk about the feasibility of testing the famous Twin Paradox, just a simulated test of course," George continued.

"Ok" the twin sons replied without hesitation.

"Let's say Mark will be the brother about to go away for a long trip in a spaceship moving at 99 percent of the speed of light, and John will be the brother to stay at home on the earth," George assigned the roles to his twin sons.

"I need some time to think over my role," Mark said. "I will also think about the role of Mark, because my role is so simple that there is no need to be worried about it," John remarked.

"Go ahead, guys," George agreed gladly.

About one month later, their discussion came back. This time Mark began first: "I've found a big problem in the famous story of the Twin Paradox. According to the idea of mass increase with speed, that is, the so-called concept of relativistic mass, my mass would be increased up to 503 kilograms at 99 percent of the speed of light from the present 71 kilograms. You know, 503 kilograms is more than 1100 pounds." "Yes, I've also seen the same problem," John agreed quickly.

The result from his twin sons was really a great surprise to George, so he recalculated. And he also got the same result as Mark and John! The great surprise, however, still continued hanging around George. He kept asking himself: why, why ... and why I have not thought of the issue of mass increase with speed, I should have noticed that earlier ... I should have thought of that before.

But an even greater surprise to George was the second result of Mark. After pausing a short while, Mark continued: "Moreover, from the idea of length contraction embedded in special relativity, my length, which is actually my front-to-back thickness, suppose I would keep upright in the spaceship and face the direction in which it is moving, would be only about 1/7.1 of my original thickness. This factor would also expand my area by the ratio of 1 to 7.1, plus the idea of mass increase with speed would enlarge my area from 1 to 7.1 either, thus my face-to-face area would be about 50 times the original area. So I would virtually become a large picture." "Oh, really?" John also felt a surprise a bit this time. But his recalculation showed that the result of Mark was correct indeed! After George checked the result of his two sons, correct though, he felt the surprise was still so big that he couldn't face it.

For a couple weeks, George kept asking and thinking: "Why I have not considered time dilation, mass increase with speed, and length contraction together, why, why, and why? I should have thought of the three things together, I should have...." In that kind of mood, poor George was not happy at all, of course. His wife, Jennifer, certainly noticed the sadness of dear George; she also knew why he was so unhappy after asking her twin sons. She decided to help her dear and poor George, but didn't know how. Eventually, or by accident, Jennifer told George the fable of 'the Blind Men and the Elephant'. After George heard this wonderful fable, being suddenly enlightened, he became happy again. This fable seemed to have an amazing effect on George!

Then what is the fable of 'the Blind Men and the Elephant'? Is it a miraculous panacea? I couldn't help sharing it with dear readers here. Once upon a time, there were six blind men who lived in a village in today's India. Every day they stood on the side of the road with begging as their living. They had often heard of elephants, but had never seen one, for being blind, how could they? One morning, it happened that an elephant was led down the road at which they exactly stood. When they heard that an elephant was passing by, they asked the drover to stop the beast so that they could have a "look". Of course they could not look at him with their eyes, but they thought they might learn what kind of animal he was by touching and feeling him. Then you should see they trusted their own sense of touch so much.

The first blind man happened to put his hand on the side of the elephant. So he said: "Well, well, this beast is exactly like a wall."

The second tightly grasped one of the elephant's tusks and felt it. So he said loudly: "You're quite mistaken. He's round, smooth, and sharp. He's more like a spear than anything else."

The third happened to grab the elephant's trunk. "Both of you are completely wrong. This elephant is right like a snake," he said.

The fourth opened both his arms and closed them around one of the elephant's legs. "Oh, how blind you are!" he cried. "It's very clear that he's round and tall like a tree."

The fifth man was very tall, so he caught one of the elephant's ears. "Even the blindest person must see that this elephant isn't like any of the things you name at all!" he said. "He's exactly like a huge fan."

The sixth man went forward for touching and feeling the elephant. He was old and slow, so it took him quite some time even to find the elephant. Eventually, he got hold of the beast's tail. "Oh, how silly you all are!" cried he. "The elephant isn't like a wall, or a spear, or a snake, or a tree; neither is he like a fan. Any person with eyes on head can see that he's exactly like a rope."

After the drover and the elephant left, the six men sat by the roadside all day, quarrelling about the elephant. They could not agree with one another, because each of them so surely believed that he knew what the beast looked like. It is not merely blind men who make such stupid mistakes. People who can see sometimes may act just as foolishly.

Having had the two stories above, let us still go back to the famous Twin Paradox and finish the discussion about this highly influential Paradox. Up to here, and at this moment, several important and noticeable questions about this Paradox are probably hovering over the heads of some,

A WONDERFUL GIFT TO THE READERS OF "A BRIEF HISTORY OF TIME"

even many, readers. Most importantly or noticeably, what bitter lesson does the Twin Paradox really teach us? And what painful lesson can we draw from this Paradox? More specifically, since the Twin Paradox turns out to be clearly a paradox once relativistic mass (mass increase with speed) comes in, why has it been so prevalent for a long time? Since the Twin Paradox turns out to be obviously a paradox only if seen together with relativistic mass and length contraction (which is used to interpret the phenomenon of length becoming shorter that appears at high speed), why has it been a fashionable example in special relativity? In other words, or more plainly, since the Twin Paradox turns out to be really and truly a paradox within the paradigm of special relativity itself (thus, one may suddenly realize that the so-called famous Twin Paradox turns out to be an 'infamous' story), why have some theoretical physicists been delighted to talk about it without boredom?

To these questions, different people may have different answers because different people can look at the same thing in quite different ways. But I like to share my answer with dear readers here (though I believe that many insightful readers may have better answers than me). In science, if several aspects are involved in the same theory, one should consider and carefully examine them together to check on whether they are compatible or not; these different aspects should be closely scrutinized simultaneously to inspect whether they are self-contradictory or not. That is, we ought to view the different aspects of the same theory as a whole, in order to make sure they are compatible or consistent with each other, in order to ensure they are not self-contradictory with one another. Otherwise, it would be highly risky—we could easily make the ridiculous mistakes like in 'the Blind Men and the Elephant', if our eyes were merely kept on one aspect while forgetting the others. Of course, it would be even more dangerous—we could easily, even inevitably, make stupid mistakes, if we simply threw away the other aspects in dealing with one aspect, once had noticed or perceived that self-contradiction is inescapable when these aspects meet together. (Commentator: the history of science development has clearly told us, from both positive and negative experiences, such a bitter or painful lesson: wrong thinking from wrong concept is the number one enemy of science! And it seems quite safe to say that all good or experienced scientists would definitely agree with and/or firmly believe: correct thinking is crucially and decisively important in science; correct concept is the key to developing correct thinking.)

Having come here, we have clearly seen two facts about special relativity. One is that special relativity turns out to be a theory that could not be really understood, because it turns out to be clearly and seriously self-contradictory, as in turn is because its two core concepts, length contraction and time dilation, are clearly and seriously self-contradictory. The other is that the famous Twin Paradox turns out to be really a paradox within the paradigm or stereotype of special relativity itself, which can enable us to see more clearly the self-contradictory feature of special relativity. These two facts are, therefore, the two pieces of hard evidence that are more than sufficient to show that special relativity turns out to be indeed a theory that is clearly and seriously self-contradictory. So, with these two facts kept in mind, it seems to be a rational or realistic estimate that one can have a good grasp of the key point of this chapter: because special relativity turns out to be clearly and seriously self-contradictory, it could not be really understood, either in essence or in fact. (Commentator: yes, such a key point is pretty clear! This is because there is such a self-evident truth or quite obvious rationale in science: no one could really understand a theory that turns out to be clearly and seriously self-contradictory. In fact, this self-evident truth is not only the fully recognized common sense or basic knowledge in science, but has also become an irrefutable or undeniable rule in the field of science. In reality, it is no exaggeration to say that all rational people would totally agree with this self-evident truth. So a clear, also inescapable, implication of this self-evident truth is: if one had known a theory turns out to be clearly and seriously self-contradictory, he or she would definitely not hold that he/she could really understand the theory. More explicitly thus more directly, *before* knowing the clear fact that special relativity turns out to be clearly and seriously self-contradictory, some people, especially some of those respected professionals in physics, might have believed they had really understood special relativity; however, *after* knowing this clear fact, they can and will clearly realize, though perhaps somewhat unexpectedly, that their earlier belief turns out to be merely an illusion.)

Up to this moment, one can clearly see that the key point of this chapter (which is that special relativity turns out to be clearly and seriously self-contradictory, thus could not be really understood) has been sufficiently shown by the two irrefutable or hard facts just summed up. Even so, I still feel it is better or necessary to go the extra mile along the direction of this key point, considering this key point itself is probably quite new to some people, especially to some respected professionals in physics. How to go the extra mile? Because special relativity has <u>five</u> main components, the task of going the extra mile is thereby to be carried out by inspecting whether these components are compatible or not in the following several paragraphs; that is, to check on whether these five components can be put into the same package of special relativity.

The five main components of special relativity are: length contraction (for interpreting length becomes shorter at high speed), time dilation (for interpreting time runs slower at high speed), relativistic mass (mass increase with speed), and its two postulates. Thus, if these five components are not compatible—that is, if they cannot be put into the same package of special relativity (which is a further demonstration of the self-contradictory feature of special relativity from the wide angle of its entire structure), the side on which special relativity could not be really understood will get an important, extra confirmation. On the other hand, if these five components are compatible—that is, if they can be put into the same package of special relativity, then the side on which special relativity could be understood will still retain a piece of important evidence. So the basis or criterion of judging the result to be obtained from carrying out this task is rational, objective, clear and fair to both sides.

As a preparation, let us have a quick look at the two postulates of special relativity. Its first postulate says that: the speed of light is the same for all observers, regardless of their motion relative to the source of light; 'the constancy of the speed of light', to be simpler. (As a reminder or comparison, this postulate was described by Hawking in his book *A Brief History of Time* as: "All observers should measure the same speed of light, no matter how fast they are moving.") The second postulate of special relativity tells us: all observers moving at constant speed should have the same physical laws. (Also as a reminder or comparison, in *A Brief History of Time* this postulate was said as: "The laws of science should be the same for all freely moving observers, no matter what their speed.")

Now we get ready to inspect whether the five main components of special relativity are compatible or not. Let us start from finding out the answer to the question: can length contraction, time dilation and relativistic mass be compatible with the second postulate of special relativity? Since length, time and mass are all three fundamental physical quantities with the *same* fundamental status, the laws of physics are eventually attributed to describing the relationships among these three fundamental quantities—other related composite quantities, like velocity, acceleration and force, are derived from them. (Please also notice that other fundamental quantities, such as temperature and electric current, are not related to the topics that special relativity involves.) This straightforward rationale thus determines that, for the goal of the second postulate of special relativity to be tenable, the *prerequisite* is that the changes in length, time and mass have to be *at the same rate* in the state of moving at a certain high speed (that is, they three either all decrease or all increase with the same rate in such a state). However, the actual meanings of length contraction, time dilation and relativistic mass are: length becomes shorter, and time runs slower, whereas mass becomes larger. Consequently, the combination of length contraction, time dilation and relativistic mass literally takes away such a *prerequisite*, the minimum or least requirement for the second postulate of special relativity to be tenable. Therefore, it turns out that length contraction, time dilation and relativistic mass cannot be compatible with the second postulate of special relativity at all.

One may argue or ask like: how about if relativistic mass stands together with the second postulate of special relativity? Let us see what would happen then. If relativistic mass (mass increase with speed—that is, mass becomes larger) were put together with this postulate, and maintained its validity, then the inevitable requirement on length and time would become that length becomes longer and time runs quicker. Such a requirement directly conflicts with length contraction and time dilation, simply because they respectively correspond to length becomes shorter and time runs slower that appear in the situation of high speed (which, in turn, is because in special relativity, length contraction is responsible for interpreting length becomes shorter, and time dilation is responsible for interpreting time runs slower). Not only that, since length contraction and time dilation are the direct consequences of fitting the first postulate of special relativity, the effect of this postulate is actually reflected in length contraction and time dilation; that is, they represent this postulate in fact. In other words, relativistic mass, when standing together with the second postulate of special relativity, cannot be compatible with its first postulate at all. (What should be clarified or pointed out is: the above analyses and conclusions neither say the two postulates of special relativity *themselves* are not correct, nor these two postulates themselves *alone* are not compatible.)

Through the information or evidence in the two paragraphs above, one can clearly see that the five main components of special relativity, which are length contraction (for interpreting length becomes shorter), time dilation (for interpreting time runs slower), relativistic mass (mass increase with speed), and its two postulates, turn out to be obviously incompatible in fact; that is, they cannot be put into the same package of special relativity at all. Undoubtedly, such an obvious incompatibility clearly shows the self-contradictory feature of special relativity from the angle of its overall structure. (Commentator: quite noticeably, this obvious incompatibility, being the third time showing the self-contradictory feature of special relativity, can substantially help one perceive this self-contradictory feature more comprehensively and thoroughly. This perception can enable one to realize this self-contradictory feature more clearly and definitely.) Thus, as a reminder, the side on which special relativity could not be really understood has gotten an extra confirmation. (At this moment, some, even many, sharp and insightful readers may think of or raise the related question like: can we find a new theory that is able to reveal the mechanism behind the two postulates of special relativity? If so, we can see more clearly the newly revealed *truth* that the five main components of special relativity are indeed not compatible. Dear readers, you will see the mechanism underlying these two postulates soon in this chapter—before the end of this chapter.)

Up to now, it becomes clear that, around the key point of this chapter (again which is: special relativity turns out to be a theory that is clearly and seriously self-contradictory, thus could not be really understood in fact), stand three sufficient reasons or three pieces of hard evidence—actually three irrefutable facts. One is that the two *core or key* concepts of special relativity, length contraction and time dilation (respectively for interpreting length becomes shorter and time runs slower at high speed), have been revealed to be clearly and seriously self-contradictory, which shows that special relativity turns out to be clearly and seriously self-contradictory. Another is that the famous Twin Paradox turns out to be really and truly a paradox within the paradigm of special relativity itself, which can enable us to realize the self-contradictory feature of special relativity even more clearly. The third reason is that the five main components of special relativity—length contraction, time dilation, relativistic mass (mass increase with speed), and its two postulates, turn out unable to be compatible; that is, they cannot stay in the same package of special relativity at all. So this reason can enable one to see further, from an extensive angle, the self-contradictory feature of the entire structure of special relativity. All in all, with these three pieces of hard evidence, the key point of this chapter is not only a plain fact, but also an explicitly shown fact; and through these three pieces of hard evidence, one can clearly see this key point from three different directions or angles, like 3-D visual effects. (Related questions and answer: it is known that special relativity has passed observational tests; can these tests change or affect this key point? If not, why? Answer: no, they cannot; because all observational tests of special relativity have neither targeted nor even intended to touch, of course, cannot change these three pieces of hard evidence that not only collectively and consistently point to, but also explicitly and sufficiently show such a plain fact: special relativity turns out to be clearly and seriously self-contradictory, thus could not be really understood. That is, these tests cannot change or affect this key point at all.)

At this moment, and up to here, it seems to be a rational estimate that many, even most, receptive readers have understood the key point of this chapter (which is that special relativity turns out to be clearly and seriously self-contradictory, thus could not be really understood, either in essence or in fact), because this key point has been clearly shown by sufficient evidence, actually by more than sufficient evidence. Even so, it seems neither unrealistic nor unreasonable that some, even many, careful readers may still hope to view this key point from new and wide angles, so that they can see this key point more clearly and explicitly, grasp it more tangibly and definitely, and realize it more comprehensively and confidently. Besides, if some people, especially some of the respected professional people in physics, are still psychologically reluctant towards this key point, these new and wide angles, via substantially and markedly expanding one's vision on this key point, can effectively remove or reduce their possible reluctance of this kind; this effect is a considerable help for them to face this key point calmly and rationally. (Question: then what are these new and wide angles? Answer: they are some clear, even manifest, thus also noticeable, clues that can be a substantial help in reminding or inspiring one to think of, or ponder, or realize: this key point turns out to be rather rational or quite normal.)

Out of the above considerations, also as an extra effort on the key point of this chapter, it seems not inappropriate to provide some of this kind of clear and/or noticeable clues for dear readers here; and I am delighted to do that. Undoubtedly, these clues must have the basic and common feature that can naturally, or easily, direct one to chew over and digest the plain *reality* that special relativity could not be really understood, for three plain reasons. This *reality* is not only definitely clear, but also inescapable, when viewed from the perspective of this key point; in fact, this *reality* is clearly displayed in this key point; moreover, this *reality* is exactly what this key point has directly told us. (Of course, dear readers will have their own judgment or final say on whether these clues are really helpful or not, truly useful or not, after knowing about the specific information of these clues, which is reflected in the three prominent, also fundamentally and crucially important, clear clues to be analyzed and discussed in the following several or seven paragraphs.)

First clue: special relativity is unable to solve the most fundamental problem in front of itself. Since the major subject of special relativity is to tell us that time runs slower and length becomes shorter in the situation of high speed (this subject has many other similar expressions or descriptions, like: both time and length change with speed; time and length are variable thus relative at different speeds; there are different time and length at different speeds, etc.), clearly, even obviously, the problem of *why* time runs slower and *why* length becomes shorter in such a situation is definitely the most fundamental problem in front of special relativity. Nevertheless, special relativity doesn't have the ability to resolve this most fundamental problem at all, because it is incapable of revealing the mechanism behind these two *whys*. For instance, while special relativity tells us that time runs slower in the situation of high speed, it is unable to tell us *why* time runs slower in such a situation.

In fact, this inability of special relativity has been perplexing many brilliant physicists for a long time—that is, this inability, being a tacitly admitted *fact* in reality, is actually a clearly perceived or explicitly realized *fact* in truth; such an inability, believe it or not (admit it or not), is also an irrefutable or undeniable *fact* of course. (Related questions and answer: wait a minute, please. Could one clearly perceive and explicitly realize, even only from the basic and prominent feature of special relativity, the *fact* that special relativity is really unable to solve the most fundamental problem in front of itself—*why* time runs slower and *why* length becomes shorter in the situation of high speed? How? Answer: yes, and definitely! This could be easily done if one views or thinks over this *fact* in the simple and clear way like the following. If special relativity had been able to solve this most fundamental problem, its first postulate (this postulate says that the speed of light is the same for all observers, regardless of their motion relative to the source of light, or simply referred to as 'the constancy of the speed of light', as just mentioned above) would not have been necessary at all. And so, the irrefutable or undeniable *actuality* that special relativity indispensably and desperately needs its first postulate can enable one to perceive this *fact* clearly and realize it explicitly; this irrefutable or undeniable *actuality* also unavoidably shows that one has no way to deny this *fact*. Not only that, if special relativity had been able to solve this most fundamental problem, its first postulate would no longer have been a postulate; along with the constant and specific reminder from such a naked *truth:* within the paradigm or stereotype of special relativity, its first postulate is always a postulate! As a result, this naked *truth* can help one perceive and realize this *fact* even more clearly and explicitly; and with this naked *truth*, one cannot deny this *fact*. More than that, the first postulate of special relativity, or its postulate of 'the constancy of the speed of light', being one of the most famous, most important and most influential postulates in physics, turns out to be actually an impartial eyewitness and unforgettable reminder of the definite existence of this *fact*, which can make or help one face this *fact* bravely and rationally, perceive it clearly and impressively, realize it explicitly and confidently.) (Related clarification or reminder: all the observational or experimental tests of special relativity don't and can't change the *fact* that special relativity is really unable to solve the most fundamental problem in front of itself, *why* time runs slower and *why* length becomes shorter in the situation of high speed. This is because there is such general knowledge or common sense in science: observational or experimental tests themselves have neither the function nor the ability to answer the questions about the *whys* or solve the problems of the *whys*, therefore the task or purpose of observational or experimental tests is not to deal with these questions or these problems at all; the task of answering

WHY DIFFICULT TO UNDERSTAND "SPACE AND TIME"

these questions or solving these problems is, or is supposed to be, responsible by theories instead.) Thus, this fundamentally important clue can substantially help one chew over, digest and realize the plain reality that special relativity couldn't be really understood—at least in a sense or from a certain angle, through the thinking like: because special relativity is a theory that doesn't have the ability to solve the most fundamental problem in front of itself, it turns out to be quite reasonable thereby pretty understandable, at least not irrational at all, that one couldn't really understand such a theory.

The above analysis can be seen more clearly from the angle of another noticeable feature of special relativity. This feature is that the starting point of special relativity is its first postulate, simply referred to as 'the constancy of the speed of light'. Please carefully notice that the speed of light is a *derived*, composite quantity, rather than a *fundamental* quantity—the speed of light is the distance it has traveled divided by the time it has taken. And more candidly thus more plainly, the starting point of special relativity is merely the tactic of "forcing" two *fundamental* quantities (time and length) to match up a *derived*, composite quantity (the speed of light). (Commentator: yes, that's true. In fact, one can safely say that all the professionals in physics are well aware of such a basic knowledge: time and length are two fundamental quantities, whereas speed or velocity, no matter whether it is the speed of a car or a plane or even a projectile, of course including the speed of light, is a derived, composite quantity.) As a result, this noticeable feature can easily enable us to think of or think over: given that the starting point of special relativity is a derived, composite quantity, rather than a fundamental one; whereas time and length (both are the core quantities that special relativity deals with) are fundamental quantities, and so, it is readily understandable, at least not a big surprise at all, that special relativity is unable to resolve its facing most fundamental problem that is exactly about time and length. Accordingly, and again, it is rather rational or quite normal that one could not really understand a theory that cannot solve the most fundamental problem in front of itself. (One can see this fundamentally important clue even more clearly, through a sharp contrast, after knowing about how to solve *this* most fundamental problem—*why* time runs slower and *why* length becomes shorter in the situation of high speed. Such information will appear soon in this chapter.)

Second clue: special relativity turns out unable to face the daily-seen *reality* of how photons, being the tiny and discrete quanta or particles of light, get their velocity c (the speed of light) from the electron that emits them. On one side, special relativity says that the speed of light is an unattainable speed for *any* object that has a mass; objects with mass must move at the speed less than that of light, no matter how to increase their speed.

According to such a statement, electrons, since having mass, would never reach the speed of light (an electron having mass is a fully recognized fact, also a universally accepted fact). On the other side, it has been clearly known that photons are emitted from electrons (in fact, virtually all the photons seen in our daily life are emitted from electrons. There are a lot of such examples: the photons from family light bulbs, the photons from portable flashlights, and the photons from vehicle lights, even the photons from the open fire of your barbecue dinner, and so on. Moreover, some of the most important theories in modern physics, established in the early 20th century, have also fully recognized that photons are emitted from electrons. For instance, the well-known quantum mechanics, developed by Schrodinger and Heisenberg in the 1920s, openly admits that photons are emitted from electrons; and the famous model of the hydrogen atom, put forward by Niels Bohr in 1913, clearly and explicitly tells us that photons are emitted from electrons. Both theories are the crucial materials in the textbooks on modern physics; that is, it is quite safe to say that most of the professional people in physics must have known these two theories very well). However, it has also been well recognized and fully accepted that the velocity of photons cannot be increased after they are emitted from electrons. Thus, one can clearly see that, through these two sides, there is a fatal gap between the reachable maximum velocity of an electron and the velocity of the photons emitted by the electron. Such a fatal gap clearly shows that special relativity is utterly unable to face the daily-seen *reality* of how photons get their velocity c, the speed of light, from the electron that emits them. That is, special relativity is unable to answer, actually it prevents from answering, the crucial question: where does the velocity of photons come from? (Here what ought to be clarified or emphasized is that Maxwell's wave theory of light, due to being a wave theory, is not applicable to this question at all, simply because this question is about photons that are the tiny, discrete particles of light.)

With this inability, it thus becomes clear enough that special relativity turns out to be flatly and obviously contradictory with the known *fact* that photons are emitted from electrons. Such a plain contradiction, like a clear and indelible mark, could be a constant reminder of that special relativity turns out to be really self-contradictory, thus couldn't be truly understood. In addition, about and around this clue, there is also a clear shadow or trace, actually a historical fact, reflected in a letter to Besso from Einstein himself, the founder of special relativity: "All these fifty years of pondering have not brought me any closer to answering the question, what are light quanta?" (Commentator: if Einstein had known how photons get their velocity c—the speed of light, from the electron that emits them, most probably he would no longer have been perplexed by the long-standing puzzle of "What are light quanta?") So this historical fact can also help us, at least in a sense or from a certain angle, chew over and digest that special relativity turns out to be self-contradictory indeed, thus couldn't be really understood. (Narrator: as to the answer to the crucially important question of how photons get their velocity c from the electron that emits them, you will see it in chapter six. This answer, acting as a sharp contrast, will enable you to see this clue much more clearly.)

Along the crucially important clue we have just discussed, some, even many, sharp readers may easily think of or further realize another related self-contradictory feature of special relativity. This feature behaves as: while the starting point of special relativity is 'the constancy of the speed of light' (its first postulate, as mentioned earlier), in reality it turns out that special relativity actually prevents photons (the tiny, discrete particles of light) from getting their velocity, which is the same thing as it prevents light from getting its speed. That is, on one side, the birth of special relativity is indispensably dependent on the speed of light; but on the other side, what is ironically or interestingly true is that special relativity does not allow light to get its speed in fact. Thus, when these two sides come together; and when these two sides are considered simultaneously, the self-contradictory feature of special relativity thereby becomes clear, even obvious, which can substantially help one ponder that special relativity turns out to be truly self-contradictory, accordingly couldn't be really understood. (More than that, the realization of such a self-contradictory feature also has other profound and very meaningful implications: it is not only definitely correct, but also a clear reflection that these readers have a crucially important method of thinking. This method is the comprehensive and connected mode of thinking, being an excellent and extremely important quality in science. So all the readers having such a realization should have earnest congratulations on themselves! Please also accept my sincere congratulations on them.)

The third clue is also related to the self-contradictory feature of special relativity. In special relativity, Einstein put three-dimension space and one-dimension time together to form one four-dimensional system, in which space and time can neither be separated nor viewed independently. Nevertheless, according to the very *reality* revealed earlier in this chapter, within the paradigm or stereotype of special relativity, a meter stick measuring length (one of the three dimensions of space) and a clock measuring time cannot be put in the same spaceship moving at high speed, because they would directly deny each other if put together. Such a *reality* is actually equivalent to saying: you cannot simultaneously measure length (thus space) and time within the paradigm of special relativity. But special relativity also tells us: length contraction and time dilation (respectively for

A WONDERFUL GIFT TO THE READERS OF "A BRIEF HISTORY OF TIME"

interpreting length becomes shorter and time runs slower at high speed) are simply the observable consequences of making measurements. Therefore, when these big aspects come together; and when these big aspects are considered at the same time, it seems not difficult that one can clearly and easily perceive a profoundly and extensively important clue that unavoidably points to: special relativity turns out to be a self-contradictory theory. Quite obviously, also rather rationally, such a perception can explicitly and considerably help one face the plain reality that special relativity couldn't be really understood. (Having seen the above clues as well as the discussions to them, some, even many, readers might have realized: these clues are really a substantial and explicit help to their chewing over, digesting and facing up to this plain reality, thereby the key point of this chapter. Again, this key point is: special relativity turns out to be a theory that is clearly and seriously self-contradictory, thus couldn't be really understood. Undoubtedly, such a realization is also the best reward for my effort to provide these clues for you and discuss them with you.)

So, up to here, we have seen that the central line of this chapter has gone along the following route. Length contraction and time dilation are factually incompatible thus essentially contradictory (a basic fact) → they are clearly and seriously self-contradictory (a solid fact) → special relativity is clearly and seriously self-contradictory (a clear fact) → it turns out that special relativity couldn't be really and truly understood (a plain reality) → it turns out that general relativity couldn't be really understood in fact (a natural consequence), because special relativity is the *crucial* basis of general relativity → it turns out to be rather rational or quite normal that one has encountered enormous difficulties in understanding the concept of space and time described in *A Brief History of Time*, because this concept not only directly comes from, but also is totally determined by special relativity and general relativity. And so, it turned out that one shouldn't have felt discouraged or disappointed for having not really understood this concept. One can also see that, the key to following this central line lies with firmly grasping the clear *fact* that special relativity turns out to be clearly and seriously self-contradictory; once one has firmly grasped this clear fact, he or she can easily follow this central line.

Commentator A: having known the clear fact that special relativity turns out to be clearly and seriously self-contradictory, it seems quite reasonable or realistic that some, even many, receptive and sharp readers might have fully realized the two inescapable implications or consequences of this clear fact. One is the above-mentioned plain reality that special relativity turns out to be a theory that couldn't be really understood in truth. Another is that special relativity turns out to be a theory that is unable to give valid or correct explanations to the greatly important phenomena it

WHY DIFFICULT TO UNDERSTAND "SPACE AND TIME"

deals with, such as time runs slower and length becomes shorter in the situation of high speed, simply because these explanations are clearly and fatally contradictory with each other, either in essence or in fact or in both. Such a realization can easily enable these readers to yearn for some kind of new theory that is not self-contradictory, because they might not be satisfied with merely knowing that special relativity is self-contradictory.

Commentator B: yes, and surely, such a yearning is not only necessary, but also very important from the standpoint of science! In science, it is, of course, important to beware of a certain theory, particularly a highly influential theory like special relativity, turns out to be clearly and seriously self-contradictory. However, it seems far more important to know about a new theory that, in comparison with special relativity, has at least two such characteristic features or basic attributes. One is not self-contradictory; another is able to provide compatible and consistent, thus also easily comprehensible, explanations for the important phenomena that special relativity is actually unable to. So if there is such a new theory, you should let us know. Otherwise, some people, especially some of the respected professionals in physics, might have the reaction like: yes, special relativity is clearly and seriously self-contradictory, thus unable to give logical or comprehensible explanations to the phenomena it has involved, but it is still much better than not having had a theory to take care of the fundamentally important phenomena that appear at high speed, such as time runs slower and length becomes shorter. (The response from the author of this book: no worry, no hurry—what you want to see or what you are concerned about is also one of the important issues I am about to deal with very soon.)

While working on this chapter, I keep thinking over the question: how could one understand this chapter as soon as possible? Or how could one be well aware the central idea of this chapter as quickly as possible? (This central idea, as mentioned earlier, is: if special relativity turns out to be a theory that is clearly and seriously self-contradictory, it seems clear enough that no one could really understand such a theory, either in essence or in fact; then it is actually not possible to imagine that one could truly understand general relativity either, simply because special relativity is the *crucial* basis of general relativity. Thus it is reasonable and fair that one has encountered great difficulties in understanding the concept of space and time described in *A Brief History of Time*, because this concept is determined by these two theories, special relativity and general relativity.) Therefore, one can clearly see that, the key to clearly and easily understanding this central idea lies with fully grasping the plain fact: special relativity turns out to be a theory that couldn't be really understood, because it turns out to be clearly and seriously self-contradictory. Once one has fully grasped this plain fact, he or she will be clearly and easily aware this central idea, which can enable him or her to understand this entire chapter as soon as possible. Moreover, please also notice that this plain fact, being an explicitly shown fact, is exactly the core point of this central idea or the key point of this chapter, as pointed out above. Thus, as soon as one has fully realized this plain fact, he or she can understand this entire chapter clearly and easily, also effectively.

Then how could one sense this plain fact more distinctly and definitely from new and noticeable angles, so that he or she could grasp it more tangibly and confidently? I believe that many, even most, of us might have such an experience: comparison is one of the best ways to understand something clearly and easily—in order to have a better and quicker understanding of something (an idea, a concept, a theory, or the conclusion to a theory, for example), very often we need a comparison or contrast, sometimes even want to have a vivid comparison or a sharp contrast. So is in our effectively understanding the plain *fact* that special relativity turns out to be clearly and seriously self-contradictory (thus couldn't be really understood). And so, primarily for the purpose of helping dear readers have a clearer understanding of this plain fact, it seems better (or it seems not inappropriate) to introduce a new theory as a concrete comparison or a sharp contrast. (Narrator: so, what dear readers will see is a new theory that is not self-contradictory, and that uncovers the secrets of *why* time runs slower and *why* length becomes shorter at high speed. With such a new theory as a sharp contrast or vivid comparison, one can see this plain fact more clearly, perceive it more distinctly and definitely, thus realize it more accurately and confidently.) In addition, concisely introducing such a new theory is also necessary for me to fulfill the several promises I made earlier in this chapter—I shall not, and dare not, break these promises; some, even many, readers may eagerly want to see these promises come true. Having said those, let us go to this newly discovered theory now; let us start from a concise preview of its two unique features that distinguish this new theory from special relativity. (Related question: wait a moment, please; has this new theory been verified or confirmed? Answer: yes. In fact, it has been verified or confirmed from several different aspects or angles—about this crucially important issue or topic, it will be explicitly mentioned subsequently in the relevant parts; therefore, the validity and reliability of this newly discovered theory are quite positive.)

Most succinctly, this new theory was developed just a few years ago (by me, the author of this book[*]). It has two unique features compared to special relativity: not being self-contradictory; showing us *why* time runs slower and *why* length becomes shorter at high speed (by revealing the mechanism behind these two *whys*). Because of these two features, this new theory is easy to understand. For instance, its theoretical core (to be seen soon) is very similar to: if you read a clock of 8-centimeter face according to the time scale on a clock of 10-centimeter face, you will find that the clock of 8-centimeter face runs slower (than the clock of 10-centimeter face. You can draw two such clocks on two transparent papers, and read them). This is because a clock of smaller face has a smaller time scale, whereas a clock of larger face has a larger time scale. (*Independent and impartial commentator: for any theory in science, there is a well-known, objective and rational criterion, also a basic principle, which is: the things that really matter lie in *what* rather than who—lie with *what* the theory deals with or talks about, instead of who developed or discovered it. In fact, such a basic principle has been fully accepted by scientific community as general knowledge or common sense in science nowadays. Therefore, it is definitely rational and certainly reasonable to conclude that all today's qualified or knowledgeable scientists must have clearly known and totally agreed with this basic principle; that is to say, it is clear, even obvious, that no experienced or eligible scientists would have objection to such a basic, completely recognized, principle. And so, it is actually, or should be, not only rather rational but also quite reasonable if one concludes that all today's experienced or qualified scientists, including physicists of course, would persist in and uphold: if this basic principle were thrown away, science would inevitably lose a rational, objective and fair criterion; the fundamental nature and basic spirit of science would be fatally damaged; science would definitely go onto a dangerous track! So this basic principle, because it can play the role quite comparable to a cornerstone or foundation stone, is actually the cornerstone or foundation stone of science. Correspondingly, in order to ensure the integrity of science through keeping its rationality, objectivity, truthfulness and reliability, there is also such a basic professional requirement for anyone to introduce a theory, no matter who developed or found it: he or she is NOT allowed to make any exaggerated descriptions about the theory, including its meanings, implications and significance. Or more specifically, he or she must only tell truth and fact; and he/she must ensure that only truth and fact are introduced or stated!) (The response from the author of this book: surely and of course, I will strictly obey this basic professional requirement in introducing this new theory!)

Yet before going to the specific information of this new theory, it is necessary to know about its theoretical foundation, including fundamental and direct parts. First, the fundamental theoretical foundation of this new theory is the newly discovered and verified law of mass doing work (discovered by me just a few years ago). So we need to have a quick look at this law. The core principle of this law is: the amount of energy in the mass of an object is measured and determined by the amount of work done by the object's mass. (Such a core principle, when viewed from the angle of comprehension, is quite comparable or similar to that the energy of a body is measured and determined by the body's ability of doing work in classical physics. Quite obviously, such a similarity is a substantial help for one to grasp this core principle easily and quickly.) Concisely, as the exact reflection of this core principle, this law (with accurate mathematical expression) shows that, when the velocity of an object is increased, the object's mass does positive work, the object thus loses the same amount of energy as that of the work done by the mass of the object from and by consuming its mass. As a result, the core point of this law is: an object's mass doing *positive* work causes a corresponding *decrease* in the object's mass; that is, when a mass does *positive* work, the mass thus *decreases*. (One can clearly and easily understand this core point via such a simple comparison in classical physics: when a body does *positive* work, the energy of the body *decreases*.) (Related question and answer: fundamentally speaking, what is one of the most significant roles of this law? Answer: it eventually answers, by showing that mass has the ability of doing work, the fundamentally important question in science, *why* does mass have energy? And this answer is explicitly reflected in the concrete and remarkable function of this law to be seen immediately.)

This concrete and remarkable function is that this law has revealed the mechanism behind the famous mass-energy equivalence equation (which is $E = mc^2$, where c is the speed of light, m is the rest mass of an object, and E is the rest energy of the object). This mechanism turns out to be: the rest energy of an object, being the total energy contained in the rest mass of the object, is equal to the maximum ability of the object's mass doing positive work (this maximum ability is equal to mc^2, the right side of this equation). So this mechanism shows that the world-famous mc^2 (in this famous equation) turns out to be not only the total energy contained in a rest mass m, but also the maximum ability or capacity of the rest mass m doing (positive) work; and this mechanism further shows that the very reason *why* the total energy contained in a rest mass m is equal to mc^2 is because the maximum ability or capacity of the rest mass m doing (positive) work is equal to mc^2. As a result, after revealing the mechanism of this famous equation, the solid existence of this mechanism is an irrefutable fact, a bit like: having found out the continent of North America is the irrefutable fact that there is this continent. More than that, the validity and reliability of this famous equation become further solid and secure, after finding out its mechanism. (Commentator: wow! The mechanism behind this famous and great equation is eventually brought to light; this mechanism finally answers the biggest *why* underlying this great equation, *why* mass has energy—because mass has the ability of doing work! Moreover, if one thinks over this mechanism for a few minutes, it seems not difficult that he or she can clearly realize: only the law of mass doing work is able to reveal the mechanism behind this great equation, believe it or not. This realization can easily enable one to be aware: the existence of this law turns out to be a clear fact, also an undeniable truth, because the existence of this famous and great equation has become a well-known fact. Thus, even if Bingcheng Zhao, the author of this book, had not discovered this law, somebody else would find it someday, sooner or later.) (Related question and answer: what is the big or essential difference *before* and *after* revealing this mechanism? Answer: before this revealing, people merely knew mass has energy, but couldn't know *why;* after this revealing, people know *why* mass has energy via knowing that mass has the ability of doing work. Accordingly, this essential difference is also a clear demonstration of the fundamental importance of this law, corresponding to the fundamentally important status of this famous and great equation in science.)

What should be mentioned or noticed is that the information in the above paragraph is also fundamentally and crucially important to the newly discovered law of mass doing work. This fundamental and crucial importance is reflected in such a fact: this law has been verified or confirmed via its revealing the mechanism of the famous mass-energy equivalence equation, because this famous equation has passed experimental tests many times since its birth, thus being a fully recognized and universally accepted fact. With this verification or confirmation, the validity and reliability of this law are thus quite positive. (So, for this verification or confirmation, this law, on behalf of me, wants to express its sincere gratitude to all the related scientists for their great contributions that have made this famous and great equation found and verified, especially to the great scientist Albert Einstein, the founder of this equation.) In other words, the new theory that shows us *why* time runs slower and *why* length becomes shorter at high speed has been verified from the angle of its fundamental theoretical foundation; and this verification can enable one to be rationally confident in this new theory. (For this verification, this new theory, still on behalf of me, expresses its genuine thanks to all the related scientists for their tremendous and immortal contributions that have made this famous and great equation found and verified, particularly to the brilliant scientist Einstein.) (Related question and answer: because the law of mass doing work has revealed the mechanism behind this famous and great equation, can this very equation itself be a good window, through which one could see this law more clearly, thus have a better understanding of this law? Answer: yes, it can; please see the concrete information in the coming paragraph.)

One could tangibly and quickly grasp the law of mass doing work if he or she views this law from the several important and easily perceptible angles as follows. Angle A, from the large perspective of the fundamental question: why does mass have energy? The answer from this law is: because mass has the ability of doing work; in fact, only this law can answer such a big question. Angle B, the famous mass-energy equivalence equation tells us that mass and energy are equivalent; and since energy can do work—being a universally accepted fact, why can't mass? More specifically, since energy has the ability of doing work—being a fully recognized fact, and since mass and energy are equivalent, it is clear, even obvious, that mass also has the ability of doing work; it is necessary and inevitable that mass has the ability of doing work. (Otherwise, the fully acknowledged concept of 'mass-energy equivalence' would lose its most fundamental, most essential, and most important implication, this concept would thus become meaningless in fact.) Angle C, one can think over this law conversely if necessary: if mass could not have the ability to do work, it would be scientifically groundless to say that mass has energy! In other words, the known fact that mass has energy has no choice but to tell us another clear fact: mass has the ability of doing work. Angle D, since mass has the ability of doing work, it becomes quite natural that, when an object's mass does *positive* work, the object's mass thereby *decreases* (one can clearly perceive and easily comprehend this point if he or she is familiar with that in classical physics, when a body does *positive* work, the available energy of the body thus *decreases*). (Commentator: when one views the law of mass doing work through these diverse visual angles, he or she can see this law more clearly from different directions, a bit like 3-D visual effects; he/she can thus grasp this law more effectively. Once one has grasped this law, he or she can clearly and easily realize that this law can lay the solid foundation for any theory based on it.)

Second, the direct theoretical foundation of this new theory is the mass consumption caused by mass doing positive work (being the simple result of the combination of the two things just mentioned above: the law of mass doing work and the mechanism of the famous mass-energy equivalence equation revealed with this law), which shows that the mass of an object *decreases* with the increase in its velocity, by revealing why and how the object's mass is being consumed due to its doing positive work. (Related question and answer: how could one easily and quickly realize, from available and related knowledge, the clear and definite existence of this direct theoretical foundation? Answer: as long as one has known or heard of the famous mass-energy equivalence equation, which is $E = mc^2$, as just said above, he or she can clearly realize the solid existence of this direct theoretical foundation, because it is inherently connected to the mechanism of this famous equation. In other words, this famous equation, which is often said as the greatest equation in the history of science, turns out to be also a great and splendid bridge leading us to this direct theoretical foundation, which, in turn, leads us to this new theory.)

Now we are ready to go to this new theory, which is the result of an application of the mass consumption caused by mass doing positive work. Since a mass meter (the equipment used to measure mass, a balance, for example) consists of different blocks with standard mass, the mass consumption is, of course, applicable to mass scale—the mass of all standard blocks is consumed or reduced according to the mass consumption, referred to as mass scale reduction. Because any unit length of a meter stick (that is, the length scale of a meter stick, such as one meter and one centimeter) is made of a certain amount of mass, clearly the mass consumption is equally applicable to length scale too—the mass between two neighboring graduations on a meter stick is consumed or reduced according to the mass consumption, referred to as length scale reduction. Because any unit time of a clock (that is, the time scale of a clock such as per minute) is composed of a certain amount of mass, clearly the mass consumption is also equally applicable to time scale—the mass between two neighboring graduations on a clock is consumed or reduced according to the mass consumption, referred to as time scale reduction. As a result, due to the effect of the mass consumption caused by mass doing positive work, the scales of mass, length and time in the situation of moving at a certain high speed are all reduced *at the same rate* with respect to the scales in the situation of not moving or moving at a lower speed (Fig. 2.4, next page). (The situation of not moving refers to that there is always no position change with respect to a third independent, stationary reference point.)

A WONDERFUL GIFT TO THE READERS OF "A BRIEF HISTORY OF TIME"

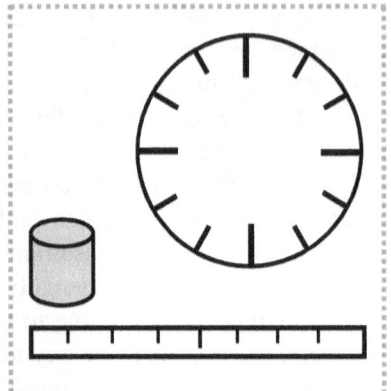 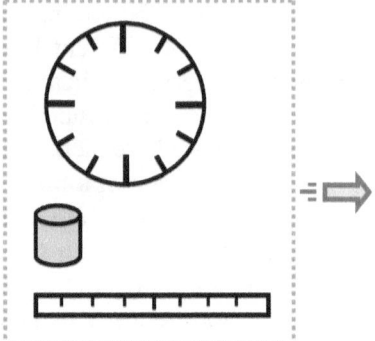

The scales of mass, length and time in the situation of *not moving or moving at a much lower speed;* that is, the mass scale, length scale and time scale in such a situation.

The scales of mass, length and time in the situation of *moving at a certain very high speed;* that is, the mass scale, length scale and time scale in such a situation.

Figure 2.4, why time runs slower in the situation of high speed (or why and how the scale of time is reduced in such a situation); why length becomes shorter in the situation of high speed (or why and how the scale of length is reduced in such a situation).

What you have seen above is the most important basic content of this new theory, also its pivotal or central content. Because it determines why and how mass, length and time are variable thus relative at different speeds by finding out the relationship in their scales; because its theoretical foundation is of mechanism-revealed nature; and because it reveals the mechanism behind its describing phenomena, this new theory has been named by me as *mechanism-revealed scales relativity theory* (MRSRT, for short). The take home message about MRSRT is: what are variable thus relative?

Answer: time and length are variable thus relative. Question: why are they variable thus relative? Answer: because their scales are variable thus relative. In addition, what should be noticed or realized is that space is also variable thus relative in MRSRT, because length scale reduction is equally applicable to width scale reduction and height scale reduction (note: length, width and height are the three dimensions that determine the size of space). (Related questions and answers: what is one of the quickest routes to go into MRSRT? Or how can one enter MRSRT easily and quickly? Answer: as long as one has thought of the famous mass-energy equivalence equation (again which is $E = mc^2$, where c is the speed of light, m is the rest mass of an object, and E is the rest energy of the object), which is often referred to as the greatest equation in the history of science, he or she can enter MRSRT easily and quickly. This is because the fundamental theoretical foundation of MRSRT, the law of mass doing work, has also revealed the mechanism of this famous and greatest equation, as just mentioned above; thus, essentially speaking or ultimately, the mechanism of this equation is this law. That is, the mechanism underlying this equation turns out to be exactly the same thing as the fundamental theoretical foundation of MRSRT. And so, MRSRT and this equation are attached onto the same thing—the law of mass doing work. Therefore, this famous and greatest equation turns out to be an eye-catching sign that directs one to go into MRSRT: once one has seen this sign, he or she will have no difficulty entering MRSRT. Questions: after entering MRSRT, how could one witness and enjoy the wonderful scenery brought to us by the highlights or main points of MRSRT? And how could these main points help one see more clearly the key point of this chapter that special relativity turns out to be clearly and seriously self-contradictory, thus couldn't be really understood? Answer: please focus on the three great and remarkable tasks MRSRT has completed, being the very subject to be introduced and discussed immediately.)

Mechanism-revealed scales relativity theory (that is, this new theory) finishes three unprecedented tasks with profound implications and historic significance: showing *why* time runs slower and *why* length becomes shorter in the situation of high speed, and revealing the mechanism behind the two postulates of special relativity (these two postulates have been briefly reviewed earlier in this chapter). Because these tasks are crucially important, let us concisely go over them one by one as follows.

First of all, and most of all, this new theory, through its time scale reduction and length scale reduction, clearly tells us *why* time runs slower and *why* length becomes shorter in the situation of high speed (still Fig. 2.4). For instance, time scale reduction tells us the reason why a moving clock runs slower than a clock at rest is: a stationary observer reads the moving, scale-reduced clock with his own larger scale. This is a bit like that, when you read a clock of 8-centimeter face according to the time scale on a clock of 10-centimeter face, you will find that the clock of 8-centimeter face runs slower (than the clock of 10-centimeter face; you can draw two such clocks on two transparent papers, and read them). Therefore, time scale reduction shows us *why* time runs slower at high speed by revealing the mechanism behind this *why*. Similarly, length scale reduction tells us the reason why a moving meter stick becomes shorter than a meter stick at rest is: a stationary observer reads the moving, scale-reduced meter stick with his own larger scale. Thus length scale reduction shows us *why* length becomes shorter at high speed by revealing the mechanism behind this *why*. (In contrast, special relativity is unable to tell us *why* time runs slower and *why* length becomes shorter at high speed, because its time dilation and length contraction are incapable of revealing the mechanism behind these two *whys*; not to mention that its time dilation and length contraction turn out to be factually incompatible thus essentially contradictory with each other. As a result, this big or crucial difference, like a tangible and sharp comparison, can substantially help one digest and grasp the plain fact, also the key point of this chapter: special relativity turns out to be clearly and seriously self-contradictory, thus couldn't be really understood.) (Narrator: moreover, unlike time dilation and length contraction, time scale reduction and length scale reduction, because of showing us *why* time runs slower and *why* length becomes shorter at high speed, are compatible thus consistent with each other. So this contrasting and noticeable difference can enable one to see this key point more clearly.)

Second, this new theory reveals the mechanism behind the first postulate of special relativity, which is simply referred to as 'the constancy of the speed of light'. Because velocity scale is the ratio of length scale to time scale; because this new theory shows that the length scale and the time scale in the situation of moving at a certain high speed are both reduced *at the same rate* with respect to the scales in the situation of not moving (or moving at a lower speed), velocity scale is thereby constant, referred to as the constancy of velocity scale. For example, let us say that in the situation of not moving, the velocity scale is L / T (L and T are respectively the length scale and time scale in this situation), then in the situation of moving at 60 percent of the speed of light, the velocity scale is $(0.8L) / (0.8T) = L / T$, thus these two velocity scales are equal to each other. This is a bit like that $(ax)/(ay) = x / y$. Therefore, velocity scale, because of being the same at each of different speeds, is the same at different speeds.

Because velocity scale is the same at different speeds, the speed of light, due to being determined with velocity scale, is always the same (at different speeds). As a result, the speed of light is always the same, regardless of observer moving or not, regardless of light source moving or not, and regardless of both observer and light source moving or not. Thus, it is the constancy of velocity scale that determines 'the constancy of the speed of light'. Therefore, the mechanism behind the first postulate of special relativity is the constancy of velocity scale (this mechanism shows that the first postulate of special relativity turns out to be a special case of the constancy of velocity scale, which can substantially help one comprehend this mechanism clearly, easily and impressively, because this postulate is so famous that many people have known it or at least heard of it). (Narrator: so, the mechanism behind the first postulate of special relativity has eventually come to light; we have finally unraveled the long-standing puzzle of *why* 'the speed of light is the same to every observer, no matter how he is moving'. And so, from now on, this puzzle is no longer a puzzle to the people who have known the answer to it.) (⁺In addition, what should be pointed out or noticed is that the accomplishment of this great and remarkable task is also fundamentally important to this new theory. This importance is clearly reflected in the fact that this new theory has been verified or confirmed through its revealing the mechanism behind the first postulate of special relativity, because this extraordinarily famous and greatly important postulate has already become a fully recognized fact in science. Quite obviously, also rather rationally, this verification or confirmation can enable one to see the validity and reliability of this new theory from a fundamentally important angle.)

Having witnessed the mechanism shown above, one can clearly realize: indeed, behind the first postulate of special relativity, there is this solid mechanism! (This is a bit like: having found out the continent of North America is the hard fact that there is this continent.) With such a clear realization, it seems quite reasonable and fair that one can naturally think of or ask the closely related question like: is special relativity able to reveal the mechanism behind this postulate? The answer to this question is as clear as day: special relativity is unable to reveal the mechanism behind this postulate; that is, special relativity is utterly unable to know the underlying reason of this postulate, because this inability, believe it or not, is simply a self-evident or actually admitted *fact*. One can clearly and easily sense and grasp this *fact* via the simple and straightforward thinking like: if the mechanism behind a postulate had been revealed, the postulate would no longer have been a postulate at all; along with the constant and specific reminder from such an unavoidable *reality:* within the paradigm or stereotype of special relativity, its first postulate is always a postulate! In other words, this inability of special relativity, being a self-evident or actually admitted *fact* in essence, is certainly an irrefutable or undeniable *fact* in truth. (Related question and answer: what is the real implication or inescapable consequence of this inability? Answer: this inability is the crux or root cause that length contraction and time dilation of special relativity turn out to be clearly and seriously self-contradictory.) So this inability of special relativity can considerably help us further ponder the plain fact, also the key point of this chapter: special relativity turns out to be a theory that is clearly and seriously self-contradictory, thus couldn't be really understood.

Third, this new theory also reveals the mechanism behind the second postulate of special relativity, which says that all observers moving at constant speed should have the same physical laws. On one side, as pointed out above, this new theory shows that the scales of mass, length and time at a certain high speed are all reduced *at the same rate* with respect to the scales in the situation of not moving or moving at a lower speed. As a result, the scale ratio of mass, length and time (the ratio of mass scale, length scale and time scale) is the same in all situations, referred to as the constant scale ratio of mass, length and time. For instance, if the scale ratio of mass, length and time is M : L : T in the situation of not moving (M, L, and T are respectively the mass scale, length scale, and time scale in this situation), then in the situation of moving at 80 percent of the speed of light, the scale ratio is (0.6M) : (0.6L) : (0.6T) = M : L : T, these two scale ratios are thus equal to each other. On the other side, because mass, length and time are all three fundamental physical quantities that have the same fundamental status, all the physical laws related to special relativity are ultimately ascribed to describing the relationships among these three quantities—other related composite quantities, such as velocity and acceleration, are derived from these three fundamental quantities. (It should be noticed that other fundamental quantities, such as temperature and electric current, are not related to the topics that special relativity works on.) So, when these two sides meet together, and are considered simultaneously, evidently or inevitably appears such a clear and definite conclusion: it is the constant scale ratio of mass, length and time that determines 'all observers moving at constant speed should have the same physical laws'. Therefore, the mechanism behind the second postulate of special relativity is the constant scale ratio of mass, length and time (this mechanism explicitly points to that the second postulate of special relativity turns out to be a direct result of the constant scale ratio of mass, length and time, or a special case that comes from this constant scale ratio, to be much simpler. Such a feature can considerably help one understand this mechanism clearly, easily and impressively, because this widely recognized, famous postulate is known to be one of the most important, most prominent, most influential and most impressive postulates in physics). (Related question and answer: what is the key to perceiving and grasping the fact that this new theory has revealed the mechanism behind the second postulate of special relativity? Answer: the *prerequisite* or the least requirement for 'all observers moving at constant speed should have the same physical laws' is the constant scale ratio of mass, length and time.)

The mechanism presented above shows that really and surely, there is this solid mechanism behind the second postulate of special relativity; that is, this very postulate has this solid mechanism indeed! In other words, the clear and definite existence of this solid mechanism turns out to be an irrefutable fact. This is quite similar to: no rational people in the world are doubtful about the existence of the continent of North America, after having found out this continent. This is also a bit like: having found out diamond beneath a certain place is the solid evidence that there is diamond beneath this place. (Commentator: yes, one can clearly see this solid mechanism, simply because it has been revealed.) (++More than that, the revealing of this solid mechanism is also crucially important to this new theory. This crucial importance is clearly demonstrated or reflected in the fact that this new theory has been verified or confirmed through its revealing the mechanism behind the second postulate of special relativity, because this well-known postulate has been fully recognized in science. Quite reasonably, also very perceptibly, this verification or confirmation can enable one to see and recognize the validity and reliability of this new theory from a crucially important angle. As a result—as an explicit and noticeable result in fact, this new theory has been verified or confirmed from three different aspects or angles altogether; through these comprehensive verifications or confirmations, the validity and reliability of this new theory have been, or could be, clearly seen from different angles.)

This solid mechanism has profound implications. One of them inevitably points to: special relativity is unable to reveal the mechanism behind its second postulate (that is, special relativity is unable to dig out the underlying reason of this postulate), simply because this inability, believe it or not, is a self-evident or actually admitted *fact*. One can clearly and easily perceive and realize this *fact* via the simple and clear thinking like: if the mechanism behind a postulate had been revealed, the postulate would no longer have been a postulate at all. In other words, the very *reality*, which is that this postulate is always a postulate within the paradigm of special relativity, is exactly the hard evidence or clear manifestation that special relativity is really unable to reveal the mechanism behind this postulate. So this inability of special relativity is indeed a self-evident or actually admitted *fact;* such an inability is, of course, also an irrefutable or undeniable *fact* in truth. More noticeably, this inability is further witnessed after the mechanism behind this postulate has been revealed with a new theory. (Related question and answer: then what are the actual implications or unavoidable consequences of this inability of special relativity? Answer: this inability is the crux or root cause of the famous Twin Paradox turning out to be really a paradox within the paradigm of special relativity itself. And this inability is also the root cause or essence of the five main components of special relativity, which are length contraction, time dilation, relativistic mass (mass increase with speed), and its two postulates, turn out unable to be compatible—that is, they cannot be put into the same package of special relativity at all. And so, this very inability can enable one to see more clearly the two solid conclusions drawn earlier in this chapter.) As a result, this inability of special relativity turns out to be a clear clue that can remind us to ponder and digest the plain fact, also the key point of this chapter: special relativity turns out to be clearly and seriously self-contradictory, thus couldn't be really understood.

All in all, a series of discussions in this chapter collectively and consistently point to the following noticeable direction. Because special relativity turns out to be clearly and seriously self-contradictory, it is actually impossible or unrealistic to imagine that one could really and truly understand it, to be rational, objective and realistic (that is, it turns out to be both reasonable and fair that one couldn't really understand special relativity). Accordingly, it is actually not possible to imagine that one could truly understand general relativity, either in essence or in fact or in both, because it is *crucially* based on special relativity. Therefore, it turns out to be rather reasonable and/or quite normal that one has encountered great difficulties in understanding the concept of space and time described in *A Brief History of Time*, simply because this concept is directly and totally determined by special relativity and general relativity. (Narrator: so, this chapter turns out to be a great relief or consolation for the people who have felt it's not easy to comprehend the concept of space and time described in *A Brief History of Time*. Some, even many, readers might have already sensed that.) (Prompter: by the way, please don't forget to mention that this chapter can also be a good consolation to the two famous theories, general relativity and quantum mechanics. Why? After noticing and realizing that this chapter has revealed the clear fact that special relativity turns out to be clearly and seriously self-contradictory, it seems pretty reasonable and fair that both general relativity and quantum mechanics can easily, though perhaps suddenly, get the clear perception or vision like the following. In comparison with the very fact that the *same* theory, special relativity, turns out to be clearly and seriously self-contradictory, it appears rather "rational" that the two *different* theories, general relativity and quantum mechanics, are seriously inconsistent with each other; that is, the serious inconsistency of these two theories looks quite "reasonable".)

WHY DIFFICULT TO UNDERSTAND "SPACE AND TIME"

Having been aware of the plain reality that special relativity turns out to be a theory that couldn't be really understood, some, even many, sharp readers may naturally ask the question like: since special relativity is the crucial foundation of general relativity, could the theory of general relativity be really understood? Or does general relativity face the same fate as special relativity? This is indeed a very insightful and wise question (though this question has been conveniently involved in this chapter, such a kind of involvement is too brief to provide enough or detailed information). The answer to this question will appear as one of the important concerns or issues in the next chapter.

CHAPTER 3

WHY DIFFICULT TO UNDERSTAND "THE EXPANDING UNIVERSE"

[The Window on This Chapter]

What does the *fact* that general relativity and quantum mechanics cannot both be correct because of their inconsistency with each other (*A Brief History of Time*, 1998 version, P. 12) tell us?

It clearly tells us: if general relativity turns out to be incorrect, it's quite normal that one has difficulty understanding the idea of <u>the expanding universe</u> described in *A Brief History of Time*, because general relativity is crucially important to this idea.

Can the measured microwave radiation be valid evidence for <u>the expanding universe</u>?

Is the Doppler effect, when it is applied to the light radiated from the observed stars far from the earth, really able to provide the valid evidence for <u>the expanding universe</u>?

Let us reveal the secrets of why it's difficult to understand <u>the expanding universe</u>! Let facts tell us why it's really difficult to understand *A Brief History of Time*.

The purpose of this chapter is to provide a great relief or wonderful relaxation for the people who have felt it's difficult to understand *A Brief History of Time*. This purpose is to be materialized, in part, through digging out the crux (root causes) of why it's hard to comprehend the idea of the expanding universe described in *that book* (chapter three, "The Expanding Universe", 1998 version). After knowing this crux, you will suddenly or amazingly, but explicitly, realize: it turns out that there are more than enough reasons to justify that this idea could not be really understood. Such an explicit realization can and will enable you to be well aware: it turns out not to be your responsibility at all that you have encountered great challenges or undefeatable difficulties in following this idea.

Clearly, in order to ensure one could really understand the idea of the expanding universe, we have to make sure this idea is correct (because in science, it is self-evident that rational people with reasonable thinking could not really understand an idea that turns out to be incorrect). In order to ensure this idea is correct, we have to make sure its theoretical foundation is correct. If its theoretical foundation turns out to be incorrect, this idea then cannot be correct in truth; thus it's definitely reasonable and fair that one could not really understand this idea in fact. (Please notice that the simple and clear chain of plain reasoning in this paragraph is one of the most important basic principles to be seen or applied in this chapter.)

With the above basic principle as a clear and definite guidance, the solid *fact*, which is that general relativity and quantum mechanics cannot both be correct because they are known to be inconsistent with each other, as clearly pointed out by Hawking in his *A Brief History of Time* (1998 version, P. 12), seems to tell us: it is a wise decision to inspect whether the theoretical foundation of the idea of the expanding universe, general relativity, is really correct or not. And the *reality* that many people have felt it's not easy to follow this idea seems to encourage us: yes, the above decision is correct. Moreover, the combination of this *fact* and this *reality* seems to warn or remind us: it becomes clear or rationally obvious that there is no valid reason for not taking action to inspect whether general relativity is really correct or not. (Commentator: in fact, even only this solid *fact* stands there, it is still clear enough that no rational people would have objection to the action of doing such an inspection. Not only that, because this solid *fact* directly comes from the well-known inconsistency of the two extraordinarily famous theories, general relativity and quantum mechanics; because all the knowledgeable and insightful professional people in physics have known this well-known inconsistency very well, it thus becomes quite rational or rather realistic if one comes to such a conclusion: all the qualified or capable professional people in physics would easily realize or totally agree that it is definitely necessary or surely sensible to inspect whether general relativity is really correct or not. And this conclusion seems to be pretty obvious thus reasonably acceptable to an overwhelming majority of the experts in physics, even if not all of them!)

But before doing this inspection, we need to be clearly aware that general relativity plays a crucial and decisive role in the idea of the expanding universe. This is determined by a set of facts. Fact one, the model of the

WHY DIFFICULT TO UNDERSTAND "THE EXPANDING UNIVERSE"

Russian scientist Alexander Friedmann (no matter how many solutions this model has), as the framework of the expanding universe, is entirely based on general relativity. In fact, Friedmann's this model is merely a product of general relativity. Fact two, the actual implication of all Stephen Hawking's work on the expanding universe is simply as: only if general relativity is correct, can the universe be expanding. Fact three, Georges LeMaitre inferred that if general relativity is right, the universe must be expanding; along with that Willem de Sitter described an expanding universe using general relativity. All in all, these facts clearly demonstrate that general relativity is crucially indispensable and decisively important to the idea of the expanding universe, because it is the theoretical basis of this idea; because it is the very theory that is responsible for interpreting this idea. (Commentator: thus, this clear demonstration not only further confirms that it is definitely necessary or surely sensible to inspect whether general relativity is really correct or not, but also shows that we cannot afford not to do such an inspection in front of the solid *fact* just mentioned above.)

Then is general relativity really correct? Or can we draw the solid conclusion that general relativity cannot be correct? In chapter one, through analyzing and discussing the well-known *fact* that general relativity and quantum mechanics are inconsistent with each other, we have clearly seen that this inconsistency is not only unavoidable, but also serious, and very serious! The real implication of this serious inconsistency is that either general relativity or quantum mechanics or both cannot be correct. (This real implication, though probably sounds "harsh" to some people, is only a specified statement of the solid *fact* that general relativity and quantum mechanics cannot both be correct because of their inconsistency with each other; again, this solid *fact* has been clearly mentioned by Hawking in his book *A Brief History of Time*, 1998 version, P. 12. Therefore, this real implication ought to be very reasonable and fair in the eyes of the people who have read *that book;* that is to say, this real implication is actually quite natural or rather normal, thus should not be a surprise at all in reality. Moreover, it is no exaggeration to say that this real implication has become, or should be, easily perceptible general knowledge or common sense in physics, because the serious inconsistency of general relativity and quantum mechanics is so well known that it has already become very important general knowledge or common sense in physics; because this real implication directly comes from and is clearly determined by this serious inconsistency. In other words, this real implication is actually, or should be, pretty obvious to all the eligible or knowledgeable professional people in physics, because they know this serious inconsistency quite well. For that reason, one can safely conclude that neither qualified nor capable professional people in physics want to deny this real implication; in fact, it seems quite reasonable and realistic that all the qualified or knowledgeable professional people in physics would totally agree with this real implication.) Of course, this serious inconsistency alone cannot surely tell us that general relativity cannot be correct, to be rational, objective and fair. (However, the explicit, fully recognized existence of this serious inconsistency, via its real implication or the solid *fact* just mentioned, has sent out such a clear signal: one should not feel too surprised or uncomfortable if general relativity turns out to be really incorrect.) Thus, we need decisive evidence to determine whether general relativity is really correct or not.

Certainly, the decisive evidence must be clear and rational, true and valid, objective and fair. If general relativity is able to solve the most fundamental problem in front of itself, we may say it can be correct; if it is not able to, we cannot say it is correct of course—that is, we have no choice but to reach the unavoidable conclusion that general relativity cannot be correct. (Commentator: yes, this decisive evidence is definitely valid and certainly sufficient, as well as absolutely necessary, because of the following self-evident, clear, even obvious, plain principle or basic rule in science. Any theory, as long as it turns out unable to solve the most fundamental problem in front of itself, cannot be fundamentally correct in fact, thus cannot be correct in truth. Moreover, to an assertion or "conclusion" that claims a theory being unable to solve the most fundamental problem in front of itself as a correct theory, this most fundamental problem will definitely say NO; the fundamental nature of science will clearly say NO; the rational people in the world will impartially say NO. Therefore, this decisive evidence is not only clearly enough, but also surely enough.) Then where does this decisive evidence exist? And more specifically thus more explicitly, what exactly is this decisive evidence?

This decisive evidence turns out to be in the major subject of general relativity. Since its major subject is to tell us that space and time are variable thus relative in a gravitational field, undoubtedly, the problem of *why* space and time are variable thus relative in such a situation is the most fundamental problem right in front of general relativity. Thus clearly enough, also surely enough, the *prerequisite* (the least or minimum requirement) for general relativity to be correct is: it must have the ability to resolve or tackle this most fundamental problem. That is, if general relativity can meet this prerequisite, it can be correct; otherwise, it cannot be correct. (Narrator: this decisive evidence thus becomes whether general relativity can satisfy this prerequisite or not.)

Then can general relativity satisfy this *prerequisite*? Let us check on it. In chapter one, we have dug out the deepest reason or root cause behind the serious inconsistency of general relativity and quantum mechanics, which is that general relativity cannot solve the problem of *why* space and time are variable thus relative in a gravitational field. For instance, while general relativity tells us that time runs slower in a gravitational field, it is unable to tell us *why* time runs slower in such a situation, simply because it is incapable of revealing the mechanism behind this *why*. (In fact, this inability has been perplexing many brilliant physicists for a long time; that is, this inability of general relativity, being a tacitly admitted *fact* in reality, is certainly an explicitly perceived or clearly noticed *fact* in truth. Such an inability, needless to say, is also an actually recognized or clearly realized *fact*, thus also being an irrefutable or undeniable *fact*.) Therefore, it turns out that really and surely, general relativity is unable to solve the most fundamental problem in front of itself; consequently, general relativity cannot satisfy this *prerequisite*. (Related questions and answer: wait a minute, please. Could one clearly perceive and explicitly realize, even only from the perspective of general relativity itself, the hard *fact* that general relativity is really unable to solve the most fundamental problem in front of itself—*why* space and time are variable thus relative in a gravitational field? How? Answer: yes, and definitely! One could clearly perceive and explicitly realize this hard *fact* if he or she views it or thinks it over in the simple and effective way like the following. If general relativity had been able to solve this most fundamental problem, its postulate of 'invariant scales of length and time', which says that the scales of length and time at different points over an entire gravitational field are the same, would not have been necessary at all. And so, the irrefutable or undeniable *reality* that general relativity indispensably and desperately necessitates this very postulate can enable one to perceive this hard *fact* clearly and realize it explicitly; this irrefutable or undeniable *reality* also unavoidably shows or inevitably points to that general relativity has no choice but to admit this hard *fact*. Of course, this hard *fact* will manifest itself even further thus become more perceptible and noticeable if one thinks of or asks the simple and clear question like: if general relativity had been able to solve this most fundamental problem, who would have looked for trouble by proposing such a totally redundant, completely unnecessary postulate?!? In other words, this postulate, being an absolutely necessary and fundamentally indispensable postulate of general relativity, turns out to be actually an impartial eyewitness and specific reminder of the clear and definite existence of this hard *fact*. So with the considerable help from such an eyewitness and reminder, one could perceive and realize this hard *fact* even more clearly and explicitly. Moreover, with such an impartial eyewitness and specific reminder, one has no way to deny this hard *fact*.)

Because the *prerequisite* for general relativity to be correct cannot be met, we have no choice but to draw the clear conclusion, actually face the plain *actuality:* general relativity turns out to be incorrect. (Commentator: yes, this conclusion is obviously valid and true! This is because there is such a clear and self-evident basic principle or rule in science: if a theory is unable to solve the most fundamental problem in front of itself, clearly, even obviously, also unavoidably and undeniably, the theory cannot be fundamentally correct; if a theory cannot be fundamentally correct, then the theory cannot be correct, either in fact or in truth or in both. That is, this basic principle explicitly and candidly tells us: it is neither true nor valid to say a theory, which is unable to solve the most fundamental problem in front of itself, is a correct theory. Moreover, if a theory, especially a highly influential theory that involves the fundamentally important topics or fields of science, which does not have the ability to solve the most fundamental problem in front of itself, were claimed as a correct theory, science would inevitably lose a rational, objective and fair criterion; the most fundamental nature and spirit of science would be fatally damaged; science would definitely be misled onto a dangerous road! So by this clear and self-evident basic principle, the above conclusion, which is that general relativity turns out to be incorrect, is definitely valid and surely solid.) (Narrator: because general relativity turns out unable to be correct; because it is the theoretical basis of the idea of the expanding universe, it turns out to be rather rational or quite understandable that one has had great difficulty understanding this idea.)

If some people, especially some of the respected professionals in physics, are not psychologically happy or ready to face the plain *actuality* that general relativity turns out to be incorrect, probably because it comes too suddenly in their eyes, they are free to choose some mild expressions. One of such expressions, for example, is that general relativity cannot be fundamentally correct, simply because it is incapable of solving the most fundamental problem in front of itself. (Narrator: because general relativity turns out unable to be fundamentally correct; because it is the theoretical foundation of the idea of the expanding universe, it is not a surprise that one has had great difficulty grasping this idea.) More leniently, if some people, particularly some of those esteemed experts on general relativity, still feel "harsh" on seeing or hearing this plain *actuality*, they can temporarily skip over the issue of whether or not general relativity is incorrect, and directly go to one of the main points of this chapter, with the straightforward and simple thinking like the following. Because the theoretical basis of the idea of the expanding universe, general relativity, turns out to be a theory that is unable to solve the most fundamental problem in front of itself, it is understandable or at least not unreasonable if one has felt it's difficult to understand this idea. Anyhow, one can clearly see that the very *fact*, which is that general relativity is unable to solve the most fundamental problem in front of itself, can enable one to be well aware of the reason why it's not easy to follow this idea from the angle of its theoretical basis. (That is to say, even if some people might be psychologically hesitant or reluctant towards the plain *actuality* that general relativity turns out to be incorrect, they could still realize, through the substantial and noticeable help of the hard *fact* that general relativity doesn't have the ability to solve the most fundamental problem in front of itself, one of the most important reasons of why it's difficult to understand the idea of the expanding universe.)

After seeing the conclusion above (again which is that general relativity turns out to be incorrect, simple because it cannot solve the most fundamental problem in front of itself), some people may be wondering, even perplexing: it is said that general relativity has passed observational tests, why is it still incorrect? Please allow me to clarify a little bit. Of course, I clearly know of the three famous observational tests of general relativity: the rotation of the long axis of Mercury's orbit, light bending around the sun, and the test of time running slower in the gravitational field of the earth by the notable Harvard tower experiment. However, none of these tests can change the hard *fact* that general relativity is unable to resolve the most fundamental problem in front of itself—*why* space and time are variable thus relative in a gravitational field, simply because none of these tests has ever targeted to deal with this *fact;* not to mention that none of these tests has even had the intention to touch this *fact* at all! That is, all these observational tests actually have nothing to do with this *fact* in truth. (Related reminder: as mentioned in chapter one, being important general knowledge or common sense in science, the task or purpose of observational or experimental tests is not to answer the questions about the *whys* or solve the problems of the *whys*, because observational or experimental tests themselves have neither the function nor the ability to do that; the task of answering these questions or solving these problems is undertaken, or is supposed to be undertaken, by theories instead. Quite obviously, such general knowledge or common sense can make or help one see this hard *fact* even more clearly and definitely.) As a matter of fact, believe it or not, it is just the solid existence of this hard *fact* that exactly reveals the accurate reason why general relativity is still incorrect, even though it has passed observational tests. (Commentator: in science, a correct theory must at least satisfy two fundamental or indispensable requirements at the same time—it must agree with observational results; and it must have the ability to solve the most fundamental problem in front of itself. Otherwise, if a theory can meet *only* one of these two requirements—for instance, the theory cannot be correct, being one of the clear and self-evident principles or rules in science. That is to say, this clear and self-evident principle or rule explicitly and unavoidably tells us: if anybody concludes that a theory cannot be correct based on the hard *fact* that the theory is unable to solve the most fundamental problem in front of itself, then this conclusion is not only certainly valid, but also definitely solid. One can easily and explicitly grasp or perceive this conclusion via the simple and clear thinking like the following. If a theory is unable to solve the most fundamental problem in front of itself, it is self-evident, obvious and undeniable that the theory cannot be fundamentally correct; if a theory cannot be fundamentally correct, the theory actually cannot be correct—either in essence or in truth or in both. In other words, it is neither correct nor true, neither rational nor valid to say a theory, which is unable to solve the most fundamental problem in front of itself, is a correct theory.)

Even with the clarification above, some dear readers might still feel it is a bit too abstract or 'empty' to conclude that general relativity cannot be correct, based on the *fact* that it cannot resolve the most fundamental problem in front of itself, which is *why* space and time are variable thus relative in a gravitational field. (Of course, it seems neither unusual nor irrational that others might feel this clarification itself is still somewhat too abstract or 'empty' either.) Some readers may even have questionings or complaints like: you keep saying that general relativity cannot solve the most fundamental problem in front of itself, how to solve this problem? Can this problem really be solvable? Is there even a way to solve this problem? I understand the possible reactions of this sort—at least in a sense: in order to have a clearly and easily perceptible understanding of the conclusion that general relativity cannot be correct, these dear readers seem to need a noticeable comparison or a sharp contrast that can explicitly help them have a good grasp of the above-mentioned *fact* that really and surely, general relativity turns out unable to solve the most fundamental problem in front of itself. (Narrator: no worry, no hurry, such a comparison or contrast will be provided in this chapter.)

So, primarily for providing such a kind of comparison or contrast, I don't think it's inappropriate to introduce a new theory that is able to solve the fundamentally important problem of *why* space and time are variable thus relative in a gravitational field. Besides, it seems reasonable and realistic, at least not irrational at all, that some insightful readers may think of or ask the related question like: can we have some kind of new theory that can not only accurately predict what general relativity has predicted, but also solve the fundamentally important problem general relativity cannot solve? Dear readers, you will see such a new theory in the later part of this chapter; I promise that (after seeing how to solve the problem of *why* space and time are variable thus relative in a gravitational field, dear readers will see more clearly, through a striking contrast or sharp comparison, that general relativity really doesn't have the ability to solve the most fundamental problem in front of itself). But at this moment, I don't want to distract the main attention of dear readers from the major subject of this chapter—digging out the crux (root causes) of why many people have felt it's hard to follow the idea of the expanding universe.

Alternatively and actually, the solid conclusion drawn above, which is that general relativity cannot be correct, is also inescapable when viewed from the angle of special relativity. In chapter two, we have clearly seen the unavoidable fact that special relativity cannot be correct, from the clear fact that special relativity turns out to be clearly and seriously self-contradictory, which is explicitly determined by the uncovered fact that its two core concepts, length contraction and time dilation (respectively for interpreting length becomes shorter and time runs slower that appear in the situation of high speed), turn out to be factually incompatible thus essentially contradictory. Then what is the inherent and essential connection between special relativity and general relativity? The answer is: general relativity is *crucially* based on special relativity, because general relativity is fundamentally and decisively dependent on the length contraction and time dilation of special relativity. Such an inherent and essential connection is sufficient to determine that general relativity cannot be correct. And so, we have come to the same conclusion from different routes; such a noticeable feature, of course, can enable one to see the solid conclusion or the revealed fact that general relativity turns out unable to be correct even more clearly. (Commentator: because the theoretical basis of the idea of the expanding universe, general relativity, turns out unable to be correct, it turns out to be quite reasonable, at least not irrational at all, that one has come across great difficulty in understanding this idea.)

While the revealed *fact* that general relativity cannot be correct is a simple and effective way for one to grasp the crux of the puzzle of why it's difficult to understand the idea of the expanding universe, it seems not unrealistic that some people (especially some of those respected experts on general relativity) might not be psychologically ready or accustomed to this fact at first. Considering that, I am about to provide several clear clues for these dear readers, which can be used either for digesting the *fact* that general relativity cannot be correct or for facing the *reality* that general relativity cannot be really understood. Either is constructive for one to go onto the track of knowing about why it's not easy to follow this idea, because general relativity is the theoretical basis of this idea. (Narrator: the *reality* that general relativity cannot be really understood is the rational or natural reflection of the revealed *fact* that general relativity cannot be correct, because in science, it is self-evident and obvious that rational people with reasonable thinking could not really understand a theory that turns out to be incorrect.)

First clue: general relativity predicts singularity, a mathematical point whose size is zero with infinite density and infinite temperature. However, believe it or not, physically the two aspects of the concept of singularity, which are *'infinite density and infinite temperature'*, are actually incompatible thus clearly self-contradictory in essence. Specifically, please carefully notice that these two aspects cannot coexist at all: infinite density denies motion, whereas infinite temperature requires a motion at a very near the speed of light. (Be careful! Be discreet! Related to singularity, has appeared a sort of opinion like: one cannot really argue with a mathematical point or theorem. Such a sort of opinion, nevertheless, cannot be really valid! Why? Clearly and undeniably, when a mathematical point or theorem is applied to physics, it has to obey the rules of physics. So this sort of opinion is merely a product of sneakily changing concepts or switching subjects, whatever the purposes or reasons; thus logically speaking, it is not even worthy to argue with the opinion of this sort! In terms of absurdity or ridiculousness, this sort of opinion is no less than looking for an excuse for an assertion that a 2-kilogram hen had produced a 5-kilogram egg!) Therefore, general relativity has resulted in something that is fatally self-contradictory, which is very helpful for one to face the reality that general relativity cannot be really understood. Moreover, the characteristic feature of this clue, when considered together with the other two closely related *facts* (general relativity and quantum mechanics are seriously inconsistent with each other; general relativity is unable to solve the most fundamental problem in front of itself), it seems reasonable to believe that one should not have big obstacles in digesting the revealed fact that general relativity cannot be correct.

Second clue: general relativity turns out to be clearly self-contradictory. This is reflected in the fact that the frequency and period of a photon, according to the results obtained from general relativity, do not always coexist (photons are the tiny and discrete quanta or particles of light, so a photon can be regarded as a ray of light from the angle of comprehension). For instance, I have found that, when the quantity $GM/(Rc^2)$ is equal to 0.5 (where M and R are respectively the mass and radius of a highly massive celestial body; G is the universal gravitational constant, and c is the speed of light), the frequency and period of a photon, which are calculated from general relativity, do not simultaneously exist at all. However, the relation between the frequency and period of a photon is a reciprocal relationship (that is, frequency times period equals one); in fact, this simple relationship is so well known that it has already become general knowledge or common sense in physics or in science. It is, therefore, quite reasonable and realistic if one concludes that many people must have known this simple relationship very well. (Commentator: the self-contradictory feature of general relativity is quite helpful for us to face the reality that general relativity cannot be really understood. This is because of such a self-evident rationale or plain reason in science: nobody can really understand a theory that turns out to be clearly self-contradictory!)

Third clue: the long-standing problem around the cosmological constant. For a long time, this problem has been widely recognized as a long-term unsolved, fundamentally important problem in physics. The trickiest situation in front of this problem is: there is a *huge* difference or discrepancy in the required value to the cosmological constant from different theories (this difference can be as great as 120 times among different theories), and there is no way to reconcile such an inexplicable great difference at all. In fact, this incomprehensible great difference has been baffling many intelligent and diligent physicists over the last several decades. No wonder a visionary, well-known physicist (a Nobel laureate in physics, and still alive today) once had such an insightful and famous comment: 'the problem of the cosmological constant is really a crisis in physics'. Since the cosmological constant is an important constant that originally embedded in general relativity, this tricky problem itself can be an important and impartial clue reminding one to perceive the revealed fact that general relativity cannot be correct—at least in a sense or from a certain angle.

All in all, the clues discussed above can considerably mitigate the surprise of some dear readers towards the revealed fact that general relativity cannot be correct, or at least can greatly reduce one's psychological obstacle in facing the reality that general relativity cannot be really understood. Either is helpful for one to know about the crux of why it's difficult to understand the idea of the expanding universe described in *A Brief History of Time* (because general relativity is the theoretical basis of this idea).

Having seen the clues presented and discussed above, some readers may want to go even further and ask the question like: are there some hints that can encourage us to chew over the revealed fact that general relativity cannot be correct or to think over the reality that general relativity cannot be really understood? The answer is: yes, there are. So what will come are two such hints; through them, one can also sense the implications pointing to the crux of the puzzle of why it's hard to follow the idea of the expanding universe (via pointing to the theoretical basis of this idea, general relativity).

One hint comes from a fundamentally important postulate of general relativity, the postulate of 'equivalence principle', which says that gravitational force has the same effect in increasing the velocity of an object as other traditional forces. What should be noticed is that this postulate is crucially indispensable to general relativity—without this postulate, there would be no general relativity at all; in fact, this postulate is the heart and soul of general relativity. And broadly speaking, this postulate is also one of the most influential and famous postulates in modern physics; if there had not been this postulate, the history of modern physics would have been quite different.

In the face of such a crucially important postulate, it seems neither unusual nor irrational if one thinks of or asks the crucial questions like: is there and/or what is the mechanism behind this postulate? Clearly, if there is the mechanism behind this postulate, this mechanism ought to be essentially and crucially important, corresponding to the fundamental and crucial status of this postulate; accordingly, knowing this mechanism is of profound and great significance. Also clearly enough, the key to showing the existence of this mechanism lies with revealing it. That is, the answer to these crucial questions hinges on whether one has revealed this mechanism. (Though my work has revealed the mechanism behind this postulate, the detailed information about this work is far beyond the depth and range of this popular science book; so only the key points regarding this mechanism will be concisely mentioned here as follows.)

In one sentence, the mechanism behind this postulate turns out to be the mass consumption caused by mass doing positive work under the action of acceleration (increase of velocity), either due to gravitational force or due to other traditional forces. (The mass consumption has been mentioned in the later part of chapter two.) The core point of this mechanism is: the mass of an object is consumed (becoming less) due to its doing positive work, when the object is accelerated, no matter whether the acceleration is caused by gravitational force or by other traditional forces. And the key to grasping this mechanism is: having revealed the mechanism behind this postulate is the hard evidence that there is this mechanism, a bit like: having found the continent of North America is the irrefutable evidence that there is this continent. On the other hand, general relativity is unable to reveal the mechanism behind this postulate, being a self-evident or actually admitted *fact*, thus also being an irrefutable or undeniable *fact*. (One can clearly and easily realize this inability and this fact via the simple and straightforward thinking like: if the mechanism behind a postulate had been revealed, the postulate would no longer have been a postulate at all; along with the specific and constant reminder from such an unavoidable *reality:* within the paradigm or stereotype of general relativity, this postulate is always a postulate!)

Clearly, even obviously, this inability is equivalent to that general relativity utterly skips over the mechanism behind this postulate, which, in turn, is equivalent to the naked *truth* that general relativity is actually unable to reflect this mechanism in all its various explanations. In front of such a naked *truth*, it seems quite reasonable and fair (at least not irrational at all) if one thinks of or asks the fundamentally important question like: what is the real implication or actual consequences of general relativity turning out unable to reflect the mechanism behind its crucially important postulate in its various explanations? The rational existence of such a fundamentally important question can undoubtedly be an important hint encouraging us to ponder the revealed fact that general relativity cannot be correct, at least in a sense or from a certain perspective.

In fact, one can perceive something more profound along the hint discussed above. Specifically, the mechanism mentioned above is the result of an application of the newly discovered and verified law of mass doing work. In other words, it is this law that has revealed the mechanism behind the postulate of 'equivalence principle', the crucial and fundamental postulate of general relativity. In such a situation, it seems quite rational that no one can rule out the possibility that some insightful readers may think of the essential question like: what is the law of mass doing work? Or more directly, does this law have profound implications? (Because this law has been briefly introduced in the later part of chapter two, only the key roles of this law will be concisely mentioned here as follows.)

The law of mass doing work does have profound implications, being explicitly reflected in its fundamentally important roles with great significance. For instance, it is this law that tells us *why* mass has energy (via showing mass has the ability of doing work); it is this law that shows us *why* mass and energy are equivalent; it is this law that reveals the conversion mechanism between mass and energy; it is also this law that demonstrates that the energy in the mass of an object is determined by the amount of work done by the object's mass. Moreover, it is this law that unveils the mechanism behind the famous mass-energy equivalence equation (which is $E = mc^2$, where c is the speed of light, m is the rest mass of an object, and E is the rest energy of the object); this equation is often referred to as the greatest equation in the history of science. Even more, it is this law that turns out to be the fundamental linkage between mass and energy. All in all, with and through these fundamentally important roles that have unprecedented functions and far-reaching, historic meanings, one can clearly realize that the law of mass doing work turns out to be indeed a fundamentally important physical law with profound implications. With such a realization, it seems not difficult that one can easily perceive: the reality, which is that general relativity is unable to reveal the mechanism behind its postulate of 'equivalence principle', is actually equivalent to that general relativity utterly skips over this law and its profound implications, making general relativity unable to reflect these profound implications in all its various explanations. Such a closely connected perception can be a considerable and explicit help in inspiring or reminding one to chew over the revealed fact that general relativity cannot be correct, from the angle of this crucially important, also fundamentally indispensable, postulate of general relativity.

Another hint is from some long-term unsolved, fundamentally important problems in physics, such as the Pioneer anomaly (about it, please see the Glossary of this book if necessary), the mystery of dark matter, and the mysterious source of gamma ray bursts. (Some dear readers, probably due to not in the area of physics or astronomy, might not be very familiar with the detailed information about the three famous problems above. So it is enough to know of such a simple and clear *fact:* these three problems have been fully recognized as the fundamentally important problems in physics and/or astronomy by scientific community; worldwide many scientists have been working on them over several decades, but neither conclusive nor decisive results have been obtained yet. In fact, these problems have actually become long-term unsolved, challenging problems within the paradigm of general relativity.) In such a situation, it seems neither unusual nor irrational if one thinks of or asks the related questions like: since general relativity is one of the most important theoretical pillars and foundations of physics, can it be the root and cradle of these long-standing problems? Can it even be the shackles and obstacles to solving these big problems? The rational existence of these questions can considerably mitigate one's surprise or greatly reduce one's psychological hesitation towards the revealed fact that general relativity cannot be correct, at least from a certain angle or to some extent. This effect, needless to say, is quite helpful for one to face this revealed fact calmly and rationally. (Commentator: in science, when the paradigm or stereotype of a big theory has incubated some, even many, long-term unsolved, fundamentally important problems, it is wise or quite reasonable, at least not irrational at all, to question whether *the* big theory is really correct or not.)

Reading up to here, especially having gotten the inspiration or seen the reminder from the clues and hints just analyzed and discussed above, some attentive readers may be wondering: since many people have read *A Brief History of Time*, could they find or perceive, from *that book itself*, some traces or shadows that can extend to the reality that it's hard to follow the idea of the expanding universe? In other words, are there some traces or shadows (in the book *A Brief History of Time*) that can help us think over the reality that it's difficult to understand this idea? The answer is: yes, they could; yes, there are—there are quite a few such examples. So let us analyze and discuss a couple of such examples as a concrete manifestation of these traces or shadows.

WHY DIFFICULT TO UNDERSTAND "THE EXPANDING UNIVERSE"

One such example is that Stephen Hawking has concluded: "...there must have been a big bang singularity provided only that general relativity is correct ...", (*A Brief History of Time*, 1998 version, P. 53). So the actual implication of this conclusion is: if general relativity turns out to be incorrect, there must *not* have been a big bang singularity, thus there must be no existence of the expanding universe; accordingly it becomes reasonable that one has had great or insurmountable difficulty understanding the expanding universe (this is because there is such a self-evident and plain, also clear, even obvious, principle or rule in science: no one could really understand something that doesn't exist. In fact, it is neither rational nor realistic to imagine that one could really understand something that doesn't exist). Moreover, this '*if* general relativity turns out to be incorrect' may not be far from 'general relativity turns out to be incorrect', because in the same book Hawking has also pointed out: general relativity and quantum mechanics cannot both be correct because they are known to be inconsistent with each other (*A Brief History of Time*, 1998 version, P. 12). Another such example is Hawking's conclusion as: "...the universe *could* have had a singularity, a big bang, if the general theory of relativity was correct", (*A Brief History of Time*, 1998 version, P. 52). Again and similarly, the real implication of this conclusion is: the universe could *not* have had a singularity, a big bang, if the general theory of relativity was incorrect (note: the general theory of relativity is general relativity). Anyhow, through the examples like these, some attentive and sharp readers could perceive some kind of traces or shadows that can be the reflection, at least in a sense, of the reality that it's not easy to follow the idea of the expanding universe.

Altogether and up to now, the above discussions have dug out the crux of the puzzle of why it's difficult to understand the idea of the expanding universe, from the perspective of general relativity, the theoretical foundation of this idea. This crux is: general relativity turns out unable to be correct. These discussions have, therefore, revealed such a clear conclusion: it turns out to be rather rational or quite normal that some, even many, people have encountered great difficulty in understanding the idea of the expanding universe described in *A Brief History of Time*, because general relativity is the theoretical foundation of this idea; because general relativity is the theory that is responsible for interpreting this idea. (Not surprisingly, this conclusion doesn't come alone: beside it stands a noticeable assistant; and its main role is to help us digest this conclusion. What's this assistant? Let us uncover the secret veil of this assistant.)

This noticeable assistant turns out to be the big bang singularity, the root or cradle of the idea of the expanding universe! Now let us turn our eyes to the three major features of the big bang singularity, because the real implications of these features can considerably help one chew over and digest the conclusion drawn from the discussions above, or can substantially consolidate this conclusion. Either role is positive for one to be further and well aware of the crux of why it's very difficult to understand the idea of the expanding universe, the central subject of this chapter.

First feature, the big bang singularity turns out to be based on a theory (general relativity) that cannot be correct. Consequently, also clearly, even obviously, the concept of the big bang singularity turns out unable to be correct. (As a result, the so-called big bang singularity actually becomes the claimed big bang singularity, believe it or not.) Accordingly, the actual implication of this feature is: it turns out to be quite normal that one has had difficulty understanding the idea of the expanding universe, since this idea originates from the big bang singularity. (There is a self-evident and clear rule in science, which is that, if an idea is based on a concept that cannot be correct, the idea cannot be correct; if a concept is based on a theory that cannot be correct, the concept cannot be correct. By this plain rule, it becomes quite understandable that one could not really understand the idea of the expanding universe, because this idea is based on the concept of the big bang singularity, whereas this concept is based on the theory of general relativity that turns out to be incorrect.)

Second feature, the claimed big bang singularity is actually a concept that is clearly and fatally self-contradictory, being determined by its definition (this feature has been mentioned above as a noticeable clue that points to: general relativity cannot be really understood). The claimed big bang singularity is defined as a mathematical point that was infinitesimally small (practically zero size) with infinite density and infinite temperature. However, physically the concept of the so-called singularity is clearly self-contradictory, simply because its two *crucial* aspects, '*infinite density and infinite temperature*', are actually incompatible thus clearly contradictory with each other: infinite density denies motion, whereas infinite temperature requires motion at infinitely high speed—infinitely close to the speed of light. So the claimed big bang singularity turned out to be a concept that is clearly and fatally self-contradictory, believe it or not (admit it or not). Such a feature is, of course, a substantial help for one to think over the reality that many people have encountered great obstacles in trying to follow the idea of the expanding universe, simply because the claimed big bang singularity is the root or cradle of this idea. (Related questions and answers: what is the relation between the two features of the claimed big bang singularity discussed up to here? Answer: the first feature is the cause or seed of the second feature; the second feature is the outcome of the first feature. Question: how could one see this second feature more clearly? Answer:

the information in the following two paragraphs could provide considerable and easily perceptible help.)

Be careful! Some dear readers might have already noticed that the claimed big bang singularity has incubated a sort of seemingly "convincing" opinion like: one should/could not really argue with a mathematical point. I believe that many sharp readers could be well aware that this sort of opinion cannot be really valid at all, because it is merely a fabrication or product of craftily changing concepts or switching subjects. Why? Clearly, even obviously, when a mathematical point is applied to physics, it must obey the rules of physics—not allow the two aspects of the same concept to be clearly self-contradictory. Moreover, mathematics never allows two clearly self-contradictory aspects to appear in the same concept; that is, it cannot be truly valid either even using 'a mathematical point' as an excuse. Besides, from the angle of the rules of logic, the science of reasoning, this sort of opinion, because of changing concepts or switching subjects, is clearly, even obviously, invalid; it is not even worthy to argue with this kind of opinion! Therefore, the opinion or excuse like 'one should/could not really argue with a mathematical point' doesn't and can't change the *reality* that the claimed big bang singularity turned out to be a concept that is clearly and fatally self-contradictory.

In fact, one can naturally or easily think of something more interesting or profound from the second feature of the claimed big bang singularity. For instance, having noticed that the two *crucial* aspects of the claimed big bang singularity were clearly incompatible thus obviously contradictory with each other, thereby couldn't coexist at all, some, even many, insightful readers may further realize: since the two indispensably crucial aspects of the concept of the claimed big bang singularity could NOT coexist at all, the assumed existence of such a singularity seems to have become a big question mark. Such a deep realization is definitely correct or surely valid! This penetrating realization, plus the fact that the claimed big bang singularity turned out to be based on a theory (general relativity) that cannot be correct, it seems clear enough that the asserted existence of the claimed big bang singularity becomes highly skeptical. (Commentator: yes, such a skeptical attitude is quite reasonable and fair! In science, when the two crucial aspects of the same concept don't and can't coexist at all, no rational people would think it's irrational to question the very existence of such a concept.) As a result, such an in-depth realization is, of course, a positive signal that can explicitly remind one to ponder the reality why many people have felt it's not easy to follow the idea of the expanding universe (because the claimed big bang singularity is the root or cradle of this idea).

Third feature, at the claimed big bang singularity general relativity itself breaks down. In other words, the concept of the claimed big bang singularity actually tells us: general relativity has created a place where general relativity itself is no longer workable. This feature is, therefore, a bit like the situation in which a driver encounters a dead end. A dead end means a driver has gone along wrong routes; the claimed big bang singularity seems to send out a noticeable signal that might point to: the theory of general relativity may be incorrect. Such a signal can considerably help one be psychologically prepared in facing the revealed fact that general relativity turns out to be incorrect. This preparation, needless to say, is very helpful for one to grasp the crux of why many people have felt it's hard to understand the idea of the expanding universe described in *A Brief History of Time* (because general relativity is the theoretical basis of this idea).

To make a long story short, the above three main features of the claimed big bang singularity converge on such a characteristic feature: it would be really weird if one had no challenge in understanding the idea of the expanding universe! With this characteristic feature, all the people, who have felt that this idea could not be really understood, should have gotten a great relief. (Not only that, what makes these people feel further relieved is that another marked relief is just around the corner!)

So far, it has become clear that the conclusion obtained from the angle of general relativity is definitely an effective route of getting to the crux or essence of the puzzle why it's difficult to follow the idea of the expanding universe described in *A Brief History of Time*. (Concisely, this conclusion is: because general relativity is the theoretical basis of the idea of the expanding universe, whereas the revealed or shown fact is that general relativity cannot be correct, thus it turned out to be rather rational or natural that one has had great difficulty understanding this idea. That is, it turned out to be quite understandable that this idea could not be really understood.) Nevertheless, it seems neither unrealistic nor uncommon that some dear readers, especially some of those knowledgeable readers, might hesitate about taking such an effective route; these readers are probably with the question or perplexity like: the measured microwave radiation has been interpreted as an important piece of evidence for the expanding universe, how to view this interpretation now? Predictably or realistically, the perplexity of this kind seems to be a big obstacle preventing some dear readers from moving towards the track of this effective route. (Then can we overcome or remove this big obstacle? The answer to this question will be the subject in the coming several paragraphs.)

The second important thing to do in checking whether the idea of the expanding universe could really be understood is to examine whether the (cosmic) microwave radiation (discovered by Arno Penzias and Robert Wilson in 1965) is valid or not as the evidence for this idea. If it turns out unable to be valid, the side on which this idea could not be really understood will get another piece of hard evidence. If it is valid, the side on which this idea could be understood will keep a good reason. So the basis or criterion of judging this check is objective, rational, clear and fair to both sides.

In order to check whether the microwave radiation is valid evidence or not, we need to be clearly aware of the *prerequisite* for it to be valid. If this prerequisite is satisfied, one can come to the conclusion that the microwave radiation is valid evidence for the idea of the expanding universe. On the contrary, if this prerequisite cannot be satisfied, then one has no alternative but to face the unavoidable conclusion that the microwave radiation cannot be the valid evidence for this idea. So the standard or rule of judging whether the microwave radiation is valid evidence or not is clear, objective and fair.

Then what is this *prerequisite?* Clearly, this prerequisite is that the assumed big bang was the *only* source of the microwave radiation measured today. (Let us temporarily put aside the issue of whether the assumed big bang is right or wrong, because that does not affect our ongoing work.) That is to say, the actual implication of this prerequisite is: no celestial bodies (also called heavenly bodies), in the vast universe, can be the sources (or can be the significant sources) of the microwave radiation at present.

This prerequisite, however, actually cannot be held at all, being determined by the known features of gamma ray bursts and ultrahigh-energy cosmic rays. (As an ongoing program for nearly half a century since the 1960s, one of the hottest research areas in physics or astrophysics is to identify the mysterious source of gamma ray bursts, that is, to find out where gamma ray bursts come from. For that reason, a large number of strict observations of gamma ray bursts have been carried out over the last several decades. Almost at the same time, observing ultrahigh-energy cosmic rays has also been well implemented as an important task, because the mysterious source of ultrahigh-energy cosmic rays has been widely recognized by scientific community as one of the most fundamental mysteries in physics or astrophysics (a branch of astronomy dealing with the physical and chemical structure of the stars, planets, etc.). Thanks to the great efforts of many scientists from different countries, we have known some important features about gamma ray bursts and ultrahigh-energy cosmic rays from the cumulative results of these long-term observations.) For instance, the observed features of gamma ray bursts have clearly told us such a fact: in the universe, there are *numerous* celestial bodies that can be the sources of gamma ray bursts today; and these celestial bodies can also be the significant or important sources of the microwave radiation measured at present. And the monitored features of ultrahigh-energy cosmic rays have evidently shown a similar fact too (that is, in the vast universe, there are *numerous* celestial bodies that can be the sources of ultrahigh-energy cosmic rays at the present time; these celestial bodies can also be the important sources of the microwave radiation measured currently). Moreover, these two facts, when viewed simultaneously, actually become further clear and noticeable, because gamma rays are one of the four common types of ultrahigh-energy cosmic rays.

These known features have shown, by revealing the two facts presented above, that it is definitely not possible to ensure that the assumed big bang was the *only* source of the microwave radiation measured these days. In other words, the *prerequisite* for the microwave radiation to be valid evidence for the expanding universe cannot be satisfied at all, simply because there is utterly no way to ensure its validity. Therefore, we have to come to the solid conclusion, actually face such an inescapable *reality:* the measured microwave radiation cannot be the valid evidence for the idea of the expanding universe! This reality reminds us that the side on which this idea could not be really understood has gotten another piece of hard evidence. (Commentator: the interpretation, which has regarded the microwave radiation as the evidence for the expanding universe, turned out to be clearly, even obviously, contrary to the basic rule of logical reasoning, believe it or not. This can be easily seen via the simple example as follows. Let us say a farmer, Colin, has lost a cow without any mark as identification. A few days later, a person finds a lost cow on a piece of grassland several dozen miles away from Colin's home. In order to have the valid conclusion that this cow must be Colin's, the person must make certain that *only* Colin has lost a cow. So the "harsh" reality we have just seen, which is that the microwave radiation turns out unable to be the valid evidence for the expanding universe, can also teach us such a bitter lesson: thinking in a clear and rational way, especially not violating the very basic rule of logical reasoning, is crucially important in science!)

In addition, what should be clearly aware is that the very prerequisite, which is that the assumed big bang was the *only* source of the microwave radiation measured today, is also the necessary requirement for interpreting the microwave radiation as the so-called accurate confirmation of Alexander Friedmann's first assumption, which says that 'the universe looks identical in whichever direction we look.' (Friedmann's this assumption is the necessary or indispensable condition for having had the model that is the framework of the expanding universe. This model has been concisely mentioned at the beginning of this chapter.) In other words, the measured microwave radiation turns out unable to be the valid evidence for confirming this assumption either; that is, Friedmann's this assumption actually has not experienced valid confirmation yet. Thus, it seems not unreasonable if one questions the validity of this assumption. Accordingly, it seems not irrational if one questions the validity of this model. And so, it becomes rather rational or quite normal if one questions the validity or correctness of the idea of the expanding universe (from the angle of this model).

Related comments or deep discussions (optional): up to this moment, dear readers have witnessed three pieces of closely related information that linger around and hover over the idea of the expanding universe. One is that the theoretical foundation of this idea, general relativity, turns out unable to be correct, as revealed earlier. Another is that the claimed big bang singularity (the origin or seed of this idea), being predicted with general relativity, turned out to be a concept that is clearly and fatally self-contradictory, because its two key and indispensable aspects, 'infinite density and infinite temperature', were clearly incompatible thus obviously contradictory with each other; that is, the two crucial aspects of this concept couldn't coexist at all, as analyzed above. Accordingly, believing that the claimed big bang singularity couldn't exist becomes much more reasonable than believing it could. The third piece is that the measured microwave radiation turns out unable to be the valid evidence for this idea, as just seen above. With these three pieces of inherently related information, it seems neither unrealistic nor unreasonable that some, even many, attentive and insightful readers may suddenly get the insight or perception like the following. Most probably, the idea of the expanding universe could be merely a mistaken coincidence or "combination": incorrect general relativity creates the actually not existed big bang singularity; and the measured microwave radiation was mistakenly interpreted as the evidence for this idea. The response from me: definitely and surely, such a perception is pretty insightful and sharp, because the probability or chance of such a mistaken coincidence could be very high. Yet the topic about this probability is beyond the range of this book; and I also hate to divert the main attention of dear readers from the central subject of this chapter—digging out the crux of the puzzle of why many people have felt it's difficult to understand the idea of the expanding universe. However, I really appreciate and admire this kind of insightful and sharp perception, because it can help these readers grasp the crux of this puzzle more effectively. So, all the readers who have gotten such a perception should be happy for themselves (I am happy for them too).

Having come here and at this moment, it seems neither unrealistic nor unusual that two typical reactions might come from dear readers. Some might have realized the crux of why it's difficult to understand the idea of the expanding universe, because they have grasped, at least largely, the two solid conclusions obtained above (which are: it turns out that the theoretical basis of this idea, general relativity, cannot be correct; the microwave radiation turns out unable to be the valid evidence for this idea). This realization is a great relief for them, if they have been perplexed for having not really understood this idea. On the other hand, others might have not fully grasped this crux yet. These dear readers may have the attitude like: they would/could readily, or easily, grasp this crux as long as one can sufficiently demonstrate or clearly show, with hard evidence, that the Doppler effect, when it is applied to the light radiated from the stars far from the earth, turns out unable to provide the valid observational evidence for this idea. Otherwise, it seems difficult for them to recognize this crux.

Such an attitude is completely understandable, because the idea of the expanding universe originally came from the application of the Doppler effect in the 1920s (due to the American astronomer Edwin Hubble. The result gotten from this application is thus referred to as Hubble's law in professional materials. That is, the Doppler effect played a leading or pioneering, also crucial or decisive, role for the birth of this idea). Therefore, in order to have a full view of the entire picture about the crux of why it's difficult to understand this idea, we cannot avoid such a crucially important question: can the Doppler effect, when it is applied to the light radiated from the observed stars in many other galaxies, really be able to provide the valid evidence for this idea? (Looking for the answer to this question will be the specific task in the following several paragraphs.)

The third important thing to do, also an unavoidable task, in checking whether the idea of the expanding universe could really be understood is to examine whether or not the Doppler effect, when it is applied to the light radiated from the stars far from the earth, is truly able to provide the valid observational evidence for this idea. If the answer is YES, then the side on which this idea could be understood will still keep this crucially important evidence. On the contrary, if the answer is NO, the side on which this idea could not be really understood will get the third piece of hard evidence by overthrowing or invalidating this crucially important evidence. So again and as always, the basis or rule of judging this check is rational, objective, clear and fair to both sides. (Let us carry out this task! It's not a difficult one.)

How to carry out this task then? Answer: a simple and effective method is from the *prerequisite* for the Doppler effect to be able to provide the valid observational evidence for the expanding universe. Then what is this *prerequisite?* Clearly, the ideal prerequisite is to ensure that the entire redshift of the light radiated from the observed stars is due to they are moving away from us. However, because the Doppler effect on its own is unable to tell apart whether a redshift of light is due to a star moving away from us or due to a gravitational field, this prerequisite thus becomes that the redshift due to gravitational fields (that is, gravitational redshift) is so small that it can be negligible, compared to the redshift due to the star moving away from us. In other words, this prerequisite is that *gravitational redshift must be insignificant!* If such a prerequisite holds, one can come to the conclusion that the Doppler effect is able to provide the valid evidence for the idea of the expanding universe; if not, one cannot draw such a conclusion of course, to be rational, objective and fair. That is, if such a prerequisite turns out unable to hold, one has no choice but to reach the unavoidable conclusion that the Doppler effect cannot provide the valid observational evidence for this idea. This task, therefore, turns into checking whether this prerequisite holds or not (that is, whether it is valid or not, true or false).

Then how to accomplish this task? Certainly, this task ought to be carried out from the present available knowledge, actually the known facts. We are going to check this prerequisite from four viewing angles. (Related question and answer: why are you going to check this prerequisite from four viewing angles, instead of simply one or two? It seems that one viewing angle is okay, and two viewing angles are enough. Answer: in order to check up on this crucial prerequisite comprehensively and carefully, for avoiding the stupid or irrational mistakes like in the fable of 'the Blind Men and the Elephant'—this fable has been mentioned in chapter two. Some attentive readers might have already noticed and will continue to see that comprehensively thinking way is one of the most important concepts or issues highlighted in this book. It seems both reasonable and safe to believe that all rational people, needless to say professional scientists, would/could naturally or easily agree that thinking comprehensively and systematically is crucially important in science: thinking in such a way or mode is definitely necessary to ensure that the obtained conclusion is objective and fair. Moreover, the history of science development has clearly told us such an irrefutable fact: only objective and fair conclusions are really valid and reliable, and only really valid and reliable conclusions can stand up to the test of time and history, especially the history of science development.)

First angle: from the fact that the sun, even the much smaller earth, can cause significant gravitational redshift. The gravitational redshift in the gravitational field of the sun was observed for the first time by a team from Princeton University in the early 1960s. And the gravitational redshift caused by the gravitational field of the earth was measured in 1962 in the famous Harvard tower experiment, also referred to as Pound-Rebka-Snider experiment. What should be emphasized is that this measured result is just the gravitational redshift that takes place only within the distance as short as about 20 meters nearby the surface of the earth; by this emphasis, I want to point out that the total gravitational redshift across the entire gravitational field of the earth is much larger. Given that the sun is just an ordinary, average-sized star that can cause significant gravitational redshift, it is clearly groundless to say that the gravitational redshift of other stars is insignificant. Given that even the much smaller earth (whose mass is far smaller than the sun; the mass of the sun is about 333, 000 times that of the earth) can also cause significant gravitational redshift, clearly, even obviously, it is groundless to say that the gravitational redshift of other stars is insignificant. Therefore, the solid evidence coming from this viewing angle is sufficient to show us: the *prerequisite* that gravitational redshift must be insignificant is not valid in fact; such a prerequisite turns out to be false; such a prerequisite cannot hold at all. Thus, one has to say NO to such a prerequisite.

Second angle: from the location of the sun and the earth in our Milky Way galaxy. The sun and the earth are largely at the edge of the Milky Way galaxy (in terms of the region occupied by an overwhelming majority of the stars in this galaxy). This feature determines that the gravitational field at the location of the sun and the earth is comparatively weaker (because the strength of the gravitational field of the Milky Way galaxy tends to decrease radially from its center: the closer to the center of the galaxy, the stronger the gravitational field; the farther from the center of the galaxy, the weaker the gravitational field). In fact, each of the gravitational fields of the observable galaxies has the similar tendency. Thus it is rather reasonable to infer that the gravitational fields at the locations where reside many observed stars (particularly those "interpreted" as the observational evidence for the expanding universe) in many other galaxies can be much stronger than at the location of the sun and the earth. Therefore, the resulting gravitational redshift can definitely be significant. In other words, the result from this viewing angle also has to say NO to the *prerequisite* that gravitational redshift must be insignificant.

Third angle: from the angle of dark matter. While the fundamental nature of dark matter has been a thorny problem in science for more than several decades, it has been known that the existence of dark matter causes two noticeably observed gravitational effects. One is that the stars in the outer regions of a galaxy orbit *much* faster than they would if there were only ordinary matter present. (Thus, the inevitable or inescapable implication of this effect is: due to the existence of dark matter, the gravitational field over the entire region of a galaxy is *much* stronger than it would be if only ordinary matter existed.) Another is that the light rays from distant stars are bent by the enormous gravity of dark matter (such a phenomenon is referred to as the gravitational lens, being one of the main, effective and simple ways to identify the existence of dark matter today). (Attention, please! The inevitable or inescapable implication of this observable effect is gravitational redshift, because of the known fact: light bending and gravitational redshift are the two inseparable aspects of the nature of light behaving in a gravitational field; this fact had been clearly shown by the results obtained from observing the behavior of light in the gravitational field of the sun in the 20th century.) Moreover, it has also been known that a huge amount of dark matter exists in our Milky Way galaxy and other galaxies: it is estimated that the amount of dark matter is about 5.5 times of that of the ordinary matter (such as the matter of stars and planets) in the universe.

Such a huge amount of dark matter can definitely cause significant gravitational redshift—through the inevitable or inescapable implications of the two noticeably observed gravitational effects just mentioned above (along with that the area significantly affected by a cluster of dark matter is often billions even trillions of times the area that is occupied by the cluster of dark matter, because there is a tremendously huge extending region radially spreading out a cluster of dark matter. That is, the gravitational fields in these significantly affected regions are *much* stronger than they would be if only ordinary matter existed. As a result, the light radiated from the stars in these significantly affected, huge extending regions must experience large gravitational redshift). In other words, the existence of dark matter explicitly tells us such a clear and definite fact: the gravitational redshift due to dark matter cannot be insignificant in truth! Therefore, the result from the angle of dark matter has no choice but to say NO to the *prerequisite* that gravitational redshift must be insignificant, because it turns out that this prerequisite cannot hold at all. In addition, having noticed the existence of such an enormous amount of dark matter in the universe, one can suddenly, but clearly and definitely, realize that a present opinion, like: the gravitational fields of other galaxies are not large enough to cause a significant effect on gravitational redshift, turns out to be utterly groundless in fact. (Commentator: whenever experts were talking about the Doppler effect as the observational evidence for the expanding universe, they should not have forgotten gravitational redshift. And whenever experts were thinking of gravitational redshift, they should not have forgotten the enormous 'contribution' from the huge amount of dark matter in the universe.)

Fourth angle: from the effects of black holes. Though the question like 'what is the fundamental nature of black holes?' has been a long-standing question in science, it has been known that the existence of a black hole, due to its highly massive feature (it's not unusual that the mass of a typical black hole can be hundreds of times that of the sun), can cause the gravitational field in the vast extending region outside it to be significantly stronger, actually much stronger, than without the black hole. (Related information: the region significantly affected by a black hole is much, much larger than that occupied by the black hole, often billions even trillions of times larger, because there is a hugely vast extending region radially spreading out from a black hole.) Moreover, it is estimated that there are *millions* of black holes in the Milky Way galaxy, one can thus reasonably infer that a much larger number of black holes exist in many other galaxies. Not only that, it has been recognized that there is a supermassive black hole at the center of the Milky Way galaxy; in fact, at the center of each observable galaxy, there is also a supermassive black hole. What should be noticed is that the mass of each supermassive black hole of this type is in the range of millions even billions of times the mass of the sun. And what more noticeable is that the existence of such a supermassive black hole can, not only significantly but also greatly, strengthen the gravitational field extensively, virtually over the entire region of the galaxy in which the very black hole resides.

Clearly, these numerous and massive black holes can definitely cause significant gravitational redshift in the vastly extending regions outside them: the light radiated from the stars in such regions must experience large gravitational redshift! In other words, the available knowledge about black holes clearly tells us: the gravitational redshift due to black holes cannot be insignificant. Consequently, also unavoidably, the result from the angle of black holes cannot help but say NO to the *prerequisite* that gravitational redshift must be insignificant, because this prerequisite turns out to be neither valid nor true in fact. In addition, having been aware of the existence of numerous and massive black holes in the universe, one can suddenly, but clearly and surely, realize that a present opinion, like: the gravitational fields of other galaxies are not large enough to cause a significant effect on gravitational redshift, turns out to be obviously groundless in truth. (Independent commentator: whenever experts were talking about the Doppler effect as the observational evidence for the expanding universe, they should not have forgotten gravitational redshift. And whenever experts were thinking of gravitational redshift, they should not have forgotten the tremendous 'contribution' from the numerous and massive black holes in the universe.)

Well, having witnessed the pieces of information from the four different viewing angles above, let us have a brief recall on what has happened in the scheme that the Doppler effect is used as the so-called observational "evidence" for the idea of the expanding universe. (Such a recall can help us see more clearly the entire picture composed of the conclusions from these different viewing angles.) On the whole, under such a scheme, all the gravitational redshift in the vast universe simply evaporates! Specifically, in the eyes of this scheme, the sun, even the much smaller earth, can cause significant and observable gravitational redshift, but no other observed stars in many other galaxies can give rise to significant gravitational redshift. In the eyes of this scheme, no locations (where reside the observed stars) in many other galaxies over the vast universe can have stronger gravitational fields than at the location of the sun and the earth, even the gravitational field at this location has been known to be comparatively weaker. In the eyes of this scheme, an enormously huge amount of dark matter in the universe cannot cause significant gravitational redshift. In the eyes of this scheme, the numerous and massive black holes in the universe cannot cause significant gravitational redshift. In short, in the eyes of this scheme, there is utterly no gravitational redshift at all in the vast universe! Who can swallow up all the gravitational redshift in the vast universe into his stomach? Nobody can, of course. (Commentator: science has nothing to fear but fear the wrong mode of thinking! Incorrect thinking mode or method is always the number one enemy of science! The tortuous history of science development has clearly shown and warned us of that.)

All in all, the solid evidence from each of the four angles discussed above clearly indicates that the *prerequisite* that gravitational redshift must be insignificant turns out unable to be valid! Moreover, the combination of all the pieces of evidence from these four angles is more than sufficient to demonstrate that the very *prerequisite*, again which is that gravitational redshift must be insignificant, turns out to be false! Therefore, such a plain *truth* unavoidably appears in front of us: the Doppler effect, when it is applied to the light radiated from the observed stars far from the earth in many other galaxies, turns out unable to provide the valid observational evidence for the idea of the expanding universe in reality. Consequently, it turns out to be rather rational and fair (or at least not irrational) that one has had great difficulty understanding this idea; in fact, it turns out to be both reasonable and realistic if one has felt that he or she could not really understand this idea, considering the pioneering and decisive role of the Doppler effect in this idea. Not only that, it seems to be a rational and realistic estimate that one would definitely not have held that he or she could really understand this idea, if he/she had known this plain *truth* earlier. (Related question and answer: considering the *reality* that the Doppler effect by itself is unable to tell apart whether a redshift of light is due to a star moving away from the earth or due to gravitational redshift—the redshift of light due to gravitational fields, so the conclusion drawn from the Doppler effect depends on how to interpret the observed results. This interpretation, in turn, hinges on the *prerequisite* that gravitational redshift must be insignificant. Then what is the real implication or inescapable consequence of this prerequisite turning out to be false? Answer: the actual existence of gravitational redshift has been mistakenly interpreted as that the observed stars are moving away from the earth!)

In this chapter, we have revealed the secrets of why it's difficult to understand the idea of the expanding universe described in *A Brief History of Time*. (Of course, after revealing these secrets, they are no longer secrets from now on.) One secret turns out to be the revealed fact that general relativity cannot be correct. This secret alone is sufficient to tell us: it turns out to be normal that one has difficulty understanding the idea of the expanding universe, because general relativity is the theoretical foundation of this idea. The second secret is that the measured microwave radiation turns out unable to be the valid evidence for the idea of the expanding universe. These two secrets thus jointly show us: it turns out to be quite normal that one has great challenges in understanding this idea. The third secret turns out to be that the Doppler effect, when it is applied to the light radiated from the observed stars far from the earth, cannot provide the valid observational evidence for the expanding universe. Therefore, the revealing of these three secrets collectively and clearly shows: it turns out to be rather rational thus quite understandable that one could not really understand the idea of the expanding universe; it turns out to be both reasonable and fair that this idea could not be really understood! All in all, this chapter has ultimately found the clear and definite answer to the big puzzle: why many people have felt it's not easy to comprehend the idea of the expanding universe described in *A Brief History of Time*. (Narrator: thus, this answer is also a great relief or consolation for all the people who have felt it's really difficult to understand *A Brief History of Time*, because this idea is a crucially important subject there.)

It seems that this chapter should be ended here, but not yet. Why? I want to present a wonderful gift to dear readers. What is the wonderful gift then? Because general relativity is the *crucial* theoretical basis of the idea of the expanding universe, clearly and surely, one of the best ways of knowing the secrets of why it's difficult to understand this idea lies in having a good understanding of the revealed or shown fact that general relativity cannot be correct. (Please also notice that I have pointed out earlier in this chapter: this revealed fact is a simple and effective way for one to grasp the crux of the puzzle of why it's difficult to comprehend this idea.) One of the quickest routes leading us to clearly and effectively understanding this revealed fact rests with explicitly and accurately grasping the *fact* that general relativity is unable to resolve the most fundamental problem in front of itself, *why* space and time are variable thus relative in a gravitational field. Then how can we enter such a quickest route as easily as possible, and as soon as possible? Most likely, many of us might have the experience that comparison or contrast is one of the best methods to understand something clearly and effectively: in order to have a good understanding of something (such as an idea, a theory, and the conclusion to a theory, for example), we often need a comparison or contrast. And sometimes we even hope to have a vivid comparison or a sharp contrast. So is in our understanding the *fact* that general relativity cannot solve the most fundamental problem in front of itself.

By what I have said above, some readers may conjecture: probably, this wonderful gift is a certain new theory that has solved the fundamentally important problem of *why* space and time are variable thus relative in a gravitational field! (Narrator: yes, this conjecture is definitely correct; dear readers will see such a new theory. After being aware of how this new theory has solved the problem of *why* space and time are variable thus relative in a gravitational field, dear readers will see more clearly, through a sharp contrast, the hard *fact* that indeed, general relativity doesn't have the ability to solve this problem.) In addition, introducing such a new theory is also the promise I made earlier in this chapter—I shall keep this promise by materializing it; and some, even many, readers may eagerly look forward to seeing this promise come true. (Of course, whether this gift is really wonderful or not, you will see it only if you open it. So let us open this gift! And let us have a quick look at its main features first.)

Let us start from briefly previewing the three main features of the new theory, which was developed and verified by me just a few years ago. (*Though I am the founder of this new theory, I still firmly believe that, if I had not found it, somebody else would/could find it someday, because it is a fundamentally and crucially important theory in science; and the development of science, particularly the development of physics and astronomy in the 21st century, can and will witness the profound implications and great significance of this newly discovered theory. Yet regardless of who has developed this theory, there is such a generally acknowledged and totally accepted basic principle, which is also an objective and rational criterion: for any theory in science, the things that really matter lie in *what* rather than who—lie with *what* the theory talks about, instead of who developed it. What should be mentioned is that this basic principle has been completely recognized and admitted by scientific community as general knowledge or common sense in science these days; thus, it is reasonable and realistic to believe that all today's scientists know this basic principle pretty well. In fact, if this basic principle were thrown away, science would inevitably lose a rational, objective and fair criterion; the most fundamental nature and spirit of science would be fatally damaged; science would definitely be misled onto a dangerous track; science would no longer be science at all! In other words, this basic principle, because it is the cornerstone of science or can play the role quite similar to a cornerstone, is fundamentally important and crucially indispensable to the development and advancement of science. Correspondingly, in order to maintain the integrity of science through ensuring its objectivity, truthfulness and reliability, there is such a self-evident rule, which is also a very basic professional requirement for anyone to introduce a theory, no matter who developed it: he or she is NOT allowed to make any exaggerated descriptions about the theory—including its meanings, implications and significance. And more specifically to be further explicit, he or she must only tell truth and fact; he/she must ensure that only truth and fact are introduced or presented. The response or promise from me: surely and of course, I will strictly obey this self-evident rule or abide by this very basic professional requirement in introducing this new theory!) Having said those, let us go to the three main features of this new theory now.

First of all and most all—first feature: this new theory solves the fundamentally important problem of *why* and how space and time are variable thus relative in a gravitational field, by revealing and determining *why* and how their scales are variable thus relative in such a situation. Such a fundamental feature, being the most prominent mark and highlight of this theory, is the heart and soul of this theory. (Commentator: this feature shows that this new theory has solved what general relativity cannot solve, thus reflecting the most fundamental and noticeable difference between these two theories, thereby providing a sharp contrast that is very helpful for one to see more clearly the *fact* that indeed, general relativity is unable to solve the most fundamental problem in front of itself. So this feature is actually a good reminder of the great importance and/or obvious necessity to introduce this new theory.)

Second feature: the predictions of this new theory accurately fit all the important observational results from different aspects. For instance, the values calculated with this theory precisely agree with the three famous observational results: the rotation of the long axis of Mercury's orbit, light bending around the sun, and the test of time running slower in the gravitational field of the earth by the Harvard tower experiment. (Commentator: such a feature answers one of the most important questions: has this theory been verified or tested? At the same time, such a feature is also a specific demonstration that this new theory, even though much simpler than general relativity, has accurately predicted what general relativity could predict.)

Third feature: this new theory is easy to comprehend. For example, its core point, being the epitome of its core concept to be presented soon, can be easily explained with the simple and clear method like the following. Let us say, initially at a certain location far away from the gravitational field of a massive celestial body, there are two exactly identical clocks, A and B; their faces are with the same size, which means they have the same time scale. Clock A then moves to a location nearby the gravitational field of the celestial body under the action of its gravity (attraction)—this motion causes the time scale of clock A to have become smaller, as revealed by the core concept to be seen immediately, whereas clock B still stays at its original location. Now, if you read clock A with the time scale on clock B, you will find that time runs slower (that is, clock A runs slower than clock B), because clock B has a larger time scale than clock A. This is a bit like: if you read a clock of 10-centimeter face according to the time scale on a clock of 15-centimeter face, you will find that the clock of 10-centimeter face runs slower (than the clock of 15-centimeter face; you can draw two such clocks on two transparent papers, and read them).

Having concisely previewed the three main features of this new theory, let us go to its core concept—the representative or spotlight of this theory. Overall, the core concept of this theory is: the space scale and time scale in a gravitational field (being referred to as the gravitational scales of space and time) are expressed with the gravitational scale contour lines of space and time, a bit like the contour lines on a topographical map (Fig. 3.1, next page). And the value of these gravitational scale contour lines is determined (calculated) by the equation of gravitational scales of space and time, the core equation of this theory.

Specifically, this core concept reveals the two basic aspects of the gravitational scales of space and time, being reflected in the key attributes of the gravitational scale contour lines of space and time (Fig. 3.1). One is that the gravitational scale contour lines of space and time in the gravitational field of a massive celestial body (the sun or the earth, for example) consist of infinite number of concentric spheres, with the mass center of the body as their common center. Another is that the value of these gravitational scale contour lines increases radially outward across different contour lines—first rapidly then slowly, somewhat like the contour lines of a normal basin, or similar to the shape of the bell part of a trumpet when it is held vertically with its bell part facing upwards. One can easily grasp or clearly visualize these two basic aspects via the following simple and clear examples. While a plane is taking off, it is passing through the different gravitational scale contour lines of space and time in the gravitational field of the earth from low to high value. While a plane is flying at a certain fixed height, it is traveling along the same gravitational scale contour line of space and time in the gravitational field of the earth. While a plane is landing, it is passing through the different gravitational scale contour lines of space and time in the gravitational field of the earth from high to low value. (Related questions and answers: what are variable thus relative in this new theory? Answer: space and time are variable thus relative. Question: why are they variable thus relative? Answer: because their scales are variable thus relative. Question: why? Answer: because the essence of the core concept of this theory shows so. Question: what is this essence then? Answer: the existence of a gravitational field causes a certain reduction in the scales of space and time in the gravitational field, along with that: the stronger the gravitational field, the larger the reduction. Question: how could one grasp the core concept of this newly developed theory clearly and easily, tangibly and impressively? Answer: please see the specific information in the coming paragraph.)

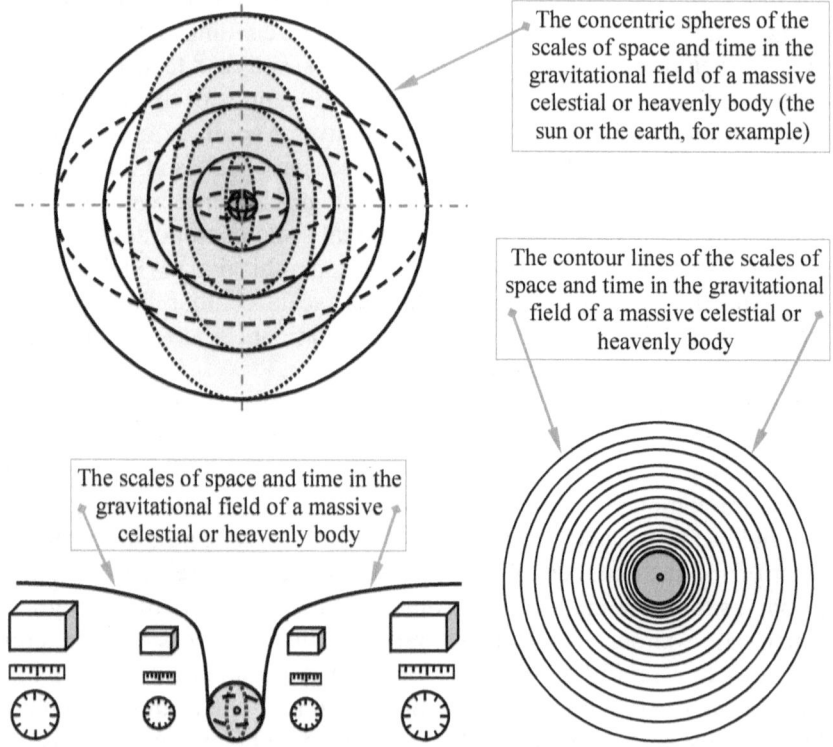

Figure 3.1, the key results from a new gravitational theory (mechanism-revealed gravitational theory). Why space and time are variable thus relative in a gravitational field (or why and how the scales of space and time are reduced in a gravitational field). So this new theory reveals the secrets or root causes of why time runs slower and why length becomes shorter in a gravitational field (in the eyes of the observers far away from the gravitational field).

WHY DIFFICULT TO UNDERSTAND "THE EXPANDING UNIVERSE"

One can tangibly grasp the core concept of this new theory via a simple and clear way like the following. Initially, at a location very far from the gravitational field of a massive celestial body, there are two exactly identical clocks—their faces are accurately with the same size of inner diameter as one meter, and they have precisely the same amount of mass; there are also two exactly identical meter sticks—their lengths are accurately at one meter, and they have precisely the same amount of mass (meter stick represents the tool of measuring length, width and height, the three dimensions that determine the size of space). And each meter stick is placed into the face of each of these two clocks, forming the shape like ⊘, thus becoming two sets of meter stick and clock, set A and set B. Then, one set of meter stick and clock, set A, is moving towards the massive celestial body under the action of its gravitational force, whereas another set of meter stick and clock, set B, still stays at the original location. In the eyes of the meter stick and clock of set B, the scales of space and time (being the length scale and time scale of set A) are gradually reduced as set A is moving towards this celestial body, because of the mass consumption from the mass of the meter stick and clock of set A, caused by their mass doing positive work (note: the mass consumption caused by mass doing positive work has been briefly introduced in the later part of chapter two). One can easily simulate and experience the above description using the simple and convenient method as follows. First, fill hydrogen gas into two same balloons to the same size, say, to the diameter of one and a half meters, and mark them as balloon A and balloon B. Then, draw a meter stick of one-meter length and a clock of one-meter face on each of these two balloons. After that, pierce a small hole with a tiny needle on balloon A; you will see that the scales of the meter stick and clock on balloon A are gradually reduced (imitating the effect of the mass consumption from the mass of the meter stick and clock of set A described above), with respect to the scales on balloon B. That is, the scales of space and time on balloon A are gradually reduced, with respect to the scales on balloon B. (Narrator: once one has grasped this core concept, he or she has actually seen the heart and soul of this new theory: it solves the fundamentally important problem of *why* and how space and time are variable thus relative in a gravitational field, by showing *why* and how the scales of space and time are variable thus relative in such a situation.)

This core concept has equipped us to go to one of the most important, also the most prominent, highlights of this new theory: *why* time runs slower and *why* length becomes shorter in a gravitational field (to the observers far away from it). Let us say, there are two observers, assigned as A and B; observer A (with clock A and meter stick A) is at a location nearby a gravitational field, whereas observer B (with clock B and meter stick B) is at a position far away from the gravitational field (Fig. 3.1, along with the description in the above paragraph). Now, if observer B reads the clock of observer A with the time scale on his own clock—clock B, he will find that time runs slower (that is, clock A runs slower than clock B), because his clock has a larger time scale than that of observer A. (Again, this is a bit like: when you read a clock of 10-centimeter face according to the time scale on a clock of 15-centimeter face, you will find that the clock of 10-centimeter face runs slower than the clock of 15-centimeter face. You can draw two such clocks on two transparent papers, and read them.) Similarly, when observer B reads the meter stick of observer A with the length scale on his own meter stick—meter stick B, he will find that length becomes shorter (that is, meter stick A becomes shorter than meter stick B), because his meter stick has a larger length scale than that of observer A. (This is a bit like the following plain principle. Let us say, you measure the distance between two cities as 800 miles on one map, whose scale is 1:500; however, if you measure this distance on another map whose scale is 1:1000, but *still using the scale of 1:500*, you will measure this distance as only 400 miles.) Therefore, this core concept clearly shows that this new theory reveals the mechanism thus the essence of: *why* and how time runs slower, *why* and how length becomes shorter in a gravitational field to the observers far away from it (because the scales of time and length are smaller in a gravitational field with respect to the scales in the region far away from the gravitational field).

After briefly introducing this new theory, we get prepared to go to the main purpose of doing that—to provide a sharp contrast or vivid comparison so that the revealed or shown *fact* that general relativity turns out unable to be correct becomes further manifest thus more noticeable; so that one could be more explicitly and accurately aware of the revealed secrets of why it's difficult to comprehend the idea of the expanding universe described in *A Brief History of Time* (because general relativity is the theoretical basis of this idea). Overall, there are four exquisite and bright windows following this newly introduced theory. Through these windows, dear readers will witness this revealed *fact* much more clearly via the simple and sharp contrasts or comparisons that come from several different directions or angles (a bit like 3-D visual effects). (In addition, what you are about to see and observe via these windows will make you experience and evaluate whether the offered gift is really wonderful or not.) So let us go over these windows one by one, specifically and concisely.

First window: having witnessed how this new theory has solved the problem of *why* space and time are variable thus relative in a gravitational field, one can clearly realize, through a sharp contrast, the hard *fact* that really, general relativity is unable to solve the most fundamental problem in front of itself, *why* space and time are variable thus relative in such a situation. Undoubtedly, such a clear realization can substantially help one perceive the revealed *fact* that general relativity cannot be correct further explicitly and definitely. This perception is one of the best ways of knowing the secrets of why it's difficult to understand the idea of the expanding universe described in *A Brief History of Time*, as pointed out earlier.

Not only that, this window also clearly displays the biggest difference, actually the most crucial or essential difference, between this new theory and the theory of general relativity. Such a difference is reflected in their different answers to the fundamentally (also crucially) important question of *why* space and time are variable thus relative in a gravitational field. The answer from this new theory is: because their scales are variable thus relative—this new theory reveals the mechanism and essence of *why* and how the scales of space and time are variable thus relative in a gravitational field. On the other hand, the answer from general relativity is: I don't know, or I can't know (because my postulate of 'invariant scales of length and time', which says that the scales of length and time at different points over an entire gravitational field are the same, prevents me from answering this *why*, even though this *why* is fundamentally important. In addition, with and through this postulate, I have honestly, though somewhat tacitly, admitted that I don't know this *why*, either in essence or in truth). So, with this biggest difference as a striking contrast or noticeable comparison, one can realize the *fact* that indeed, general relativity is incapable of solving the most fundamental problem in front of itself further vividly and profoundly. Such a vivid and profound realization is, of course, a substantial help for one to be more clearly aware of the revealed or shown *fact* that general relativity cannot be correct; such a clear awareness is one of the shortest routes of getting to the secrets of why it's difficult to understand the idea of the expanding universe said in *A Brief History of Time*. (Related question and answer: what is the root cause that results in this most crucial or essential difference between these two theories—this new theory and the theory of general relativity? Or where does this biggest difference come from? Answer: because these two theories are fundamentally different in their theoretical foundations, being the subject to be seen through the next window.)

Second window: through this window, one can clearly see the fundamental difference, also the deep difference, between this new theory and general relativity from the angle of their theoretical foundations. The direct theoretical foundation of this new theory is mechanism-revealed scales relativity theory (which has been briefly introduced in the later part of chapter two) that shows *why* time runs slower and *why* length becomes shorter at high speed by revealing the mechanism behind these two *whys*. As a result, this new theory, because its theoretical foundation is of mechanism-revealed nature, solves the fundamentally important problem of *why* and how space and time are variable thus relative in a gravitational field, by revealing the mechanism of *why* and how the scales of space and time are variable thus relative due to the existence of the gravitational field. So this new theory has been named (by me) as mechanism-revealed gravitational theory (MRGT, for short). (Related question and answer: what is one of the easiest or shortest routes to go into MRGT? Or how could one enter MRGT easily and quickly? Answer: as long as one has known or heard of the famous mass-energy equivalence equation (that is, $E = mc^2$, where c is the speed of light, m is the rest mass of an object, and E is the rest energy of the object), which is often said as the greatest equation in the history of science, he or she could easily enter MRGT. This is because the fundamental theoretical basis of MRGT (which is the law of mass doing work, having been briefly introduced in chapter two) has also revealed the mechanism of this famous and greatest equation, as mentioned in the later part of chapter two; thus essentially speaking, the mechanism of this equation is ultimately this law. As a result, the mechanism underlying this equation turns out to be exactly the same thing as the fundamental theoretical basis of MRGT; that is, MRGT and this equation are attached onto the same thing—the law of mass doing work. Therefore, this famous and greatest equation turns out to be a noticeable sign that guides one to go into MRGT directly, easily and quickly. In other words, with the substantial and explicit help from this famous and greatest equation, one will have no difficulty entering MRGT.)

On the other hand, general relativity is crucially based on special relativity that has two such basic features. First feature: as discussed and pointed out in chapter two, special relativity doesn't have the ability to solve the most fundamental problem in front of itself, *why* time runs slower and *why* length becomes shorter in the situation of high speed, because it is incapable of revealing the mechanism behind these two *whys*. For instance, while special relativity tells us that time runs slower in the situation of high speed, it is unable to tell us *why* time runs slower in such a situation. In fact, this inability of special relativity, because it has been perplexing many brilliant physicists for a long time, is either a tacitly admitted *fact* or an explicitly perceived *fact* in reality, thus also being an irrefutable or undeniable *fact* of course. (Related reminder: as to the practically very important and helpful questions: could one clearly perceive and explicitly realize, even only from the basic and prominent feature of special relativity or from its famous first postulate, the *fact* that special relativity is really unable to solve the most fundamental problem in front of itself—*why* time runs slower and *why* length becomes shorter in the situation of high speed? How? The clear answer to these questions has appeared in the middle of chapter two.) Second feature: as uncovered in chapter two and as mentioned earlier in this chapter, special relativity turns out to be clearly and seriously self-contradictory, because its two core or key concepts, length contraction and time dilation (respectively for interpreting length becomes shorter and time runs slower that appear in the situation of high speed), turn out to be factually incompatible thus essentially contradictory with each other—they would directly deny each other if met together.

Moreover, general relativity has a fundamentally important postulate, the postulate of 'equivalence principle', which says that gravitational force has the same effect in increasing the velocity of an object as other traditional forces. (Narrator: what should be well aware is that this postulate is known to be the heart and soul of general relativity. Not only that, this postulate is the indispensable foundation of general relativity: without this postulate, there would be no general relativity at all. Therefore, this postulate is crucially and decisively important to general relativity.) However, it should be noticed that such a crucially and decisively important postulate has two basic features too. First feature: (as analyzed and pointed out earlier in this chapter) as the outcome of general relativity being unable to reveal the mechanism behind this postulate, it turns out that general relativity utterly skips over this mechanism, which makes general relativity incapable of reflecting this mechanism in all its various explanations. Second feature: the reality, which is that general relativity doesn't have the ability to reveal the mechanism behind this postulate, turns out to be the root cause or "prime culprit" of general relativity being unable to solve the most fundamental problem in front of itself—*why* space and time are variable thus relative in a gravitational field.

With the pieces of information unfolded above as a sharp contrast, one can clearly realize: what this window displays is the fundamental, also essential, difference between this new theory and general relativity from the perspective of their theoretical foundations. Such a clear realization, in turn, can tangibly help one chew over and digest the revealed or shown fact that general relativity cannot be correct, thus being quickly onto one of the shortest roads of getting to the secrets of why it's difficult to understand the idea of the expanding universe described in *A Brief History of Time*, because general relativity is the theoretical basis of this idea, as pointed out earlier.

Third window: this window is a display of how the essence of this new theory drastically shakes the fundamental foundation of general relativity (this foundation is its indispensable postulate of 'invariant scales of length and time', which claims that the scales of length and time at different points over an entire gravitational field are the same). This essence, as presented above, is that the existence of a gravitational field causes a certain reduction in the scales of space and time in the gravitational field, along with that: the stronger the gravitational field, the greater the reduction. And this essence is the fundamental reason that this new theory has solved the fundamentally important problem of *why* and how space and time are variable thus relative in a gravitational field. Therefore, through the window of this essence, one can distinctly perceive thus clearly realize: this postulate of general relativity turns out to be the direct cause that prevents general relativity from solving the most fundamental problem in front of itself—*why* space and time are variable thus relative in a gravitational field. This clear realization can be a considerable help for one to be more clearly aware of the revealed fact that general relativity cannot be correct, which is one of the best ways of knowing the secrets of why it's hard to understand the idea of the expanding universe said in *A Brief History of Time* (because general relativity is the theoretical basis of this idea, as just mentioned above).

Fourth window: this window is a bit like a telescope—by which, one can clearly see or distinctly observe the far-reaching, also crucial and marked, difference between this new theory and general relativity. This difference is specifically and clearly reflected in the day-and-night contrast when they face the two greatest puzzles in science—dark matter and dark energy. This new theory is the key to unlocking the puzzles of dark matter and dark energy (because my work based on this new theory has clearly shown that). By contrast, general relativity is not only unable to solve the problems of dark matter and dark energy, but has also become the shackles and obstacles to solving them. (Unfortunately or regrettably, also very sympathetically, it seems that not many related scientists have realized this harsh reality yet; a bit like that most related or mainstream physicists had not realized that the action of looking for the "ether" turned out to be a dead alley in the second half of the 19th century. The claimed "ether" was then believed as the medium propagating light waves.) In other words, dark matter and dark energy, which are the two long-term unsolved, fundamentally important problems in physics within the paradigm of general relativity, standing there as the impartial eyewitnesses, turn out to be the positive indication that general relativity cannot be correct. (One can perceive or realize this indication via the thinking like: if general relativity were correct, the problems of dark matter and dark energy would have already been solved much earlier.) And this positive indication can be quite helpful in directing or reminding us onto one of the shortest roads of getting to the secrets of why it's difficult to comprehend the idea of the expanding universe described in *A Brief History of Time* (along the route that points to: general relativity is the theoretical basis of this idea).

Now I shall close this chapter after presenting this wonderful gift to dear readers. In fact, it seems quite noticeable that some, even many, readers, after reading this chapter, might have already acquired another better and bigger gift: they might have eventually realized that they shouldn't have felt discouraged or disappointed at all for having not really understood the idea of the expanding universe described in *A Brief History of Time*. Such an eventual realization is also one of the most wonderful and cherished gifts to me (so I have also gotten a satisfactory gift, it is quite fair, isn't it?).

CHAPTER 4

WHY DIFFICULT TO UNDERSTAND THE BLACK HOLES DESCRIBED BY HAWKING

[The Window on This Chapter]

We have been clearly and definitely told such a basic *fact:* general relativity and quantum mechanics cannot both be correct because they are known to be inconsistent with each other (*A Brief History of Time*, Stephen Hawking, 1998 version, P. 12).

What is the actual or real implication of this basic *fact* then? Either general relativity or quantum mechanics or both cannot be correct!

If general relativity turns out to be incorrect, can the black holes described by Stephen Hawking, since they are based on general relativity, be correct?

If the black holes described by Hawking cannot be correct, then could people really understand them?

What are the concrete difficulties that prevent one from really understanding the black holes described by Hawking? Could we overcome these difficulties?

The answers to these questions are about to spread out to you, to me, and to all of us!

If someone told you that a 2-kilogram hen had produced a 5-kilogram egg, you would definitely not believe it, because you know that the "news" of this sort is clearly wrong. Moreover, you are well aware that this kind of "news" cannot be understood at all—of course, you would not spend time and effort trying to understand it; and you are also pretty sure that no rational people can really understand the statement like that. Thus, in order to ensure one could really understand something (a statement or an idea, for example), we must make sure the something is truly correct, being a self-evident and clear principle in science, also a quite plain principle when one reads popular science books.

Undoubtedly, this principle should be completely applicable to the idea of the black holes described by Hawking in his popular science book *A Brief History of Time:* in order to ensure one could really understand this idea, we have to make sure this idea is correct. However, no one can *surely* say that this idea is correct in reality. Why? The theoretical basis of this idea is general relativity, whereas general relativity and quantum mechanics cannot both be correct, because these two theories are known to be inconsistent with each other, as explicitly pointed out by Hawking in the same book (*A Brief History of Time*, 1998 version, P. 12). In such a situation, it seems that we cannot avoid the crucially important questions like: is this idea truly correct? And how to check whether this idea is truly correct or not?

In science, very often the first thing (also one of the most important things) to do in checking whether an idea is truly correct is to examine whether the theoretical basis of the idea is correct or not. If its theoretical basis turns out to be incorrect, then the idea cannot be truly correct, thus no one could really understand the idea (because it is clear, even obvious, also self-evident, that rational people with reasonable thinking could not really understand an idea that turns out to be incorrect). So, in order to check whether the idea of the black holes (described by Hawking in *A Brief History of Time*) is truly correct or not, we need to inspect whether the theory of general relativity is correct or not, because it is the theoretical basis of this idea. (Narrator: some dear readers may still be shocked by the action to inspect whether general relativity is correct or not. In fact, one should not feel a surprise to this action. Why? Please remember that Hawking has clearly mentioned such a basic or solid *fact* in the same book: general relativity and quantum mechanics cannot both be correct because they are known to be inconsistent with each other—*A Brief History of Time*, 1998 version, P. 12. Thus, it is the explicit existence of this solid *fact* that asks for, or at least reminds, this inspection!) But before doing this inspection, we need to be familiar with such a clear and definite *fact:* the theory of general relativity is crucially and decisively important to the idea of the black holes described in *A Brief History of Time!*

In one sentence, to this idea the utmost importance of general relativity cannot be overestimated! For about one century since it had come to the world, general relativity has been the theoretical basis of almost all kinds of the representative work about black holes. The work done by the German astronomer Karl Schwarzschild is based on general relativity soon after this theory was developed. The result of Robert Oppenheimer is also based on general relativity. The work of Stephen Hawking and Roger Penrose is based on general relativity too. The basic feature of non-rotating black holes discovered by Werner Israel is based on general relativity again. The discovery of the features of rotating black holes by Roy Kerr is based on general relativity once again. All in all, these facts clearly show us: general relativity is the root or "mother" of black holes; and black holes are incubated from general relativity; without general relativity, the idea of the black holes (described in *A Brief History of Time*) would become water without a source, or a tree without roots. So if general relativity turned out to be incorrect, the idea of the black holes (said in *A Brief History of Time*) could not be correct in truth; it would be rather rational thus quite understandable that this idea could not be really understood in fact! Thus, the focus now is on the questions like: is general relativity really correct? And can we draw the solid or surely valid conclusion that general relativity cannot be correct?

The above questions naturally guide us or easily remind us to think of what we have seen in the last chapter (chapter three). In that chapter, we have clearly seen the revealed or shown *fact* that general relativity cannot be correct, being sufficiently determined by either or any of the following two realities.

One reality is such an inability: general relativity is unable to solve the most fundamental problem in front of itself, *why* space and time are variable thus relative in a gravitational field, simply because it is incapable of revealing the mechanism behind this *why* (related reminder: as to the practically very important and helpful questions: could one clearly perceive and explicitly realize, even merely from the perspective of general relativity, the *fact* that general relativity is really unable to solve the most fundamental problem in front of itself—*why* space and time are variable thus relative in a gravitational field? How? The clear and definite answer to them has been presented in the early part of chapter three). Moreover, this inability of general relativity can be seen even more clearly through a sharp contrast, *after* knowing how to solve this most fundamental problem with a new theory, which has been briefly introduced in the later part of chapter three. Not only that, this inability of general relativity, as analyzed and concluded in chapter one, turns out to be also the deepest reason or root cause of the serious inconsistency between general relativity and quantum mechanics, which clearly and surely indicates that this serious inconsistency, being a fully recognized, well-known *fact* in physics, is actually an impartial witness or constant reminder of this inability. Quite obviously, such a positive indication can considerably help one face this inability calmly and rationally.

This inability of general relativity has no choice but to make one draw the clear, also unavoidable, conclusion: general relativity cannot be correct. (Commentator: yes, such a conclusion is not only clear and obvious, but also rational and objective! This is because there is such a self-evident principle or rule in science: any theory, as long as it turns out unable to solve the most fundamental problem in front of itself, cannot be fundamentally correct in fact, thus cannot be correct in truth. That is to say, this self-evident principle has no alternative but to tell us that it is neither true nor objective to say a theory being unable to solve the most fundamental problem in front of itself is correct. By this self-evident principle, if one draws such a conclusion, which says that a theory cannot be correct based on the plain truth that the theory is unable to solve the most fundamental problem in front of itself, this conclusion is not only clear and inescapable, but also valid and solid.) (If some people, especially some of those respected experts on general relativity, are not psychologically happy or comfortable towards the conclusion that general relativity cannot be correct, they can choose a mild expression for sounding less harsh, such as general relativity cannot be fundamentally correct because it is unable to solve the most fundamental problem in front of itself. However, if a theory cannot be fundamentally correct, the theory actually cannot be correct in essence, believe it or not.)

Another reality comes from that general relativity is crucially based on special relativity. This is clearly reflected in the fact that general relativity is fundamentally and decisively dependent on the two core concepts of special relativity: length contraction and time dilation (respectively for interpreting length becomes shorter and time runs slower that appear at high speed). However, in chapter two, we have clearly seen the unavoidable fact that special relativity cannot be correct, from the clear fact that special relativity turns out to be clearly and seriously self-contradictory, which is explicitly determined by the uncovered fact that its two core concepts, length contraction and time dilation, turn out to be factually incompatible thus essentially contradictory—they would directly deny each other if met together. Since general relativity is crucially based on special relativity that turns out to be clearly and seriously self-contradictory, accordingly general relativity cannot be correct; since general relativity is crucially based on special relativity that turns out to be incorrect, consequently general relativity cannot be correct. Therefore, from the angle of special relativity, we have also reached the clear and solid conclusion: general relativity cannot be correct. (Commentator: yes, neither real physical experts nor the knowledgeable professional people in physics want to deny the hard *fact* that general relativity is crucially based on special relativity, because they know this hard *fact* very well. Clearly, even obviously, also plainly and undeniably, any theory in science, as long as it crucially depends on another theory that turns out to be clearly and seriously self-contradictory, thus unable to be correct, cannot be correct either.)

All in all, either or any of the two realities above is sufficient to determine and demonstrate the clear and solid conclusion that general relativity cannot be correct. Of course, these two realities together are more than enough to show this clear and solid conclusion. (Narrator: the conclusion that general relativity cannot be correct may surprise some dear readers, especially some of those respected professional people in physics who have been deeply ingrained into the paradigm or stereotype of general relativity. However, please notice that this conclusion should not be a surprise in fact, because it is merely a specific or frank confirmation of the solid *fact* that Hawking has clearly mentioned in his book *A Brief History of Time:* general relativity and quantum mechanics cannot both be correct because they are known to be inconsistent with each other—*A Brief History of Time*, 1998 version, P. 12. Clearly, even obviously, the explicit existence of this solid fact can substantially remove or reduce one's surprise towards the conclusion that general relativity cannot be correct. Such an effect is definitely a considerable help for one to face this conclusion calmly and rationally, particularly after this conclusion has been clearly shown by sufficient hard evidence, actually by more than sufficient hard evidence.)

This clear and solid conclusion, in turn, reveals such a plain fact: because general relativity is the theoretical basis of the idea of the black holes described in *A Brief History of Time*, this idea cannot be correct; accordingly it turns out to be rather rational or quite normal if one has felt it's very difficult to understand this idea; in fact, it turns out to be reasonable and fair that one could not really understand this idea. (Commentator: yes; this plain fact is so clear and obvious that it is quite noticeable to most people, if not everyone, because of the following simple and clear reason. In science, the very basic feature of an idea that cannot be correct is that *the* idea could not be really understood, because it is not only self-evident but also obvious that rational people with reasonable thinking could not really understand an idea that turns out unable to be correct. In addition, it seems quite reasonable or realistic that many readers might have such a rational attitude: even some people claimed that they could understand an idea that turns out to be incorrect, the so-called "understanding" of this sort is actually meaningless or pointless in science. This rational attitude seems to be reasonably acceptable to most people, even if not all people.)

How to digest and grasp this plain fact quickly? Clearly, the answer to this question lies with being well aware of the clear and solid conclusion that general relativity cannot be correct. If some dear readers have not yet been psychologically prepared to face this conclusion, there are several clear clues that can provide substantial help for you. First, in front of the well-known, indisputable *fact* that general relativity and quantum mechanics are seriously inconsistent with each other, no one can surely say that general relativity is correct. In fact, Hawking has clearly mentioned such a solid fact in his *A Brief History of Time:* general relativity and quantum mechanics cannot both be correct because of their inconsistency with each other (1998 version, P. 12). Quite obviously, also rather rationally, the clear and undeniable existence of this solid fact can substantially help one be psychologically prepared to face this conclusion. Second, while general relativity tells us that time runs slower in a gravitational field, it is unable to tell us *why* time runs slower in such a situation, because it is incapable of revealing the *mechanism* behind this *why*. In fact, this inability has been perplexing many brilliant physicists for a long time; that is, this inability, being a tacitly admitted *fact* in reality, is certainly an explicitly realized or clearly perceived *fact*, thus also being an actually recognized or undeniable *fact* in truth. And so, the explicit existence of this inability can definitely help one face this conclusion calmly and rationally. Third, in front of some long-term unsolved, fundamentally important problems in physics, such as the mystery of dark matter and the Pioneer anomaly (this famous anomaly has been concisely explained in the Glossary of this book), it seems rather rational, at least not irrational at all, if one thinks of or raises the related questions like: since general relativity is one of the most important theoretical pillars and foundations of physics, can it be the root and cradle of these long-standing problems? And can it even be the shackles and obstacles to solving these big problems? Quite reasonably, the rational existence of these related questions can considerably mitigate one's psychological obstacle in facing the conclusion that general relativity cannot be correct, at least from a certain angle or in a sense. All in all, the existence of these clear clues can substantially remove or reduce one's psychological hesitation or reluctance towards this conclusion, which can be very helpful for one to digest and grasp this plain fact as quickly as possible (because this plain fact is simply an inevitable and obvious outcome or product of this very conclusion).

While the plain fact revealed above has clearly told us the result obtained from checking whether the idea of the black holes (described in *A Brief History of Time*) could really be understood from the direction of its theoretical basis, general relativity, we have to admit that such a check is not complete. That is, it cannot enable us to see the *overall* picture on the issue of whether this idea could really be understood or not—for example, we have not yet known whether these black holes themselves could really be understood. (Commentator: yes, that is correct! This is because we have not yet checked on this idea from the perspective of the characteristic feature of these black holes. Undoubtedly, this characteristic feature is completely reflected or embodied in their components; that is to say, it is the components of these black holes that totally determine and represent their characteristic feature. Therefore, to inspect whether these black holes could really be understood is to check on whether their components could really be understood. If their components could be really understood, one can say these black holes could be really understood; on the contrary, if their components could not be really understood, we have no choice but to face the unavoidable *reality* that these black holes could not be really understood.) And so, in order to have a *comprehensive* picture about whether this idea could really be understood, it is necessary to check on whether the components of these black holes could really be understood or not, being the task to start immediately.

The second important thing to do in checking whether the idea of the black holes (described in *A Brief History of Time*) could really be understood is to check whether the components of these black holes could really be understood. Specifically, we need to check on whether or not these components contain some key concept that is clearly and seriously self-contradictory. If YES, then the side on which this idea could not be really understood will get another piece of hard evidence. (Commentator: yes, that is definitely valid! This is because there is such a self-evident rationale or reason in science: no one could really understand an idea that contains a key concept that turns out to be clearly and seriously self-contradictory. In fact, this self-evident rationale is so plain or obvious that it has become an irrefutable basic principle in science.) If NO, then the side on which this idea could be understood will still keep an important reason. So the basis or criterion of judging or arbitrating this check is rational, objective, clear and fair to both sides.

Then what are the components of these black holes? According to the idea of the black holes described by Hawking in his *A Brief History of Time*, a black hole has, and only has, two components. One is singularity, which is a product of general relativity as the consequence of its space-time curvature; another is the event horizon, which is the boundary of a black hole. These two components are also the characteristic feature of the black holes described in *A Brief History of Time*. Thus, let us carry out this check from the concept of singularity first; that is, let us examine whether singularity is a clearly and seriously self-contradictory concept or not, to be exact. (The work of checking on the concept of the event horizon will begin after that.)

As a consequence of the space-time curvature of general relativity, singularity is defined as a mathematical point whose size is zero with infinite density and infinite temperature. This definition actually determines and demonstrates that the most obvious or noticeable feature of singularity, also its first feature, is: the concept of singularity is factually incompatible thus essentially self-contradictory. Why? Please closely notice that, physically the two key aspects of singularity, *infinite density and infinite temperature*, cannot coexist at all: infinite density denies motion, whereas infinite temperature requires a motion at nearly the speed of light. Therefore, we have to draw the solid conclusion (or admit the simple fact) that the concept of singularity is actually clearly and seriously self-contradictory, either in essence or in truth or in both. (The fable about spear and shield, which has been presented in chapter two, could be very helpful to one's viewing and pondering the self-contradictory feature of the concept of singularity.)

As the consequence that the concept of singularity turns out to be clearly and seriously self-contradictory, believe it or not (admit it or not), it turns out to be quite normal or reasonable that the idea of the black holes described in *A Brief History of Time* could not be really understood, because this idea says a black hole contains a singularity; not to mention that the theoretical basis of singularity, general relativity, turns out to be incorrect, as just reviewed and mentioned above. (Commentator: clearly, even obviously, it is not possible to imagine that one could really understand a concept that turns out to be clearly and seriously self-contradictory, being a self-evident rationale or irrefutable truth in science. Moreover, this irrefutable truth cannot be changed by any argument or sophistry, being the topic to be coming.)

What should be pointed out is that, if one wants to be well aware that the concept of singularity turns out to be, in fact, clearly and seriously self-contradictory, perhaps he or she needs to get rid of the hindrance or interference that arises from any kind of argument or sophistry trying to defend this concept. Some readers might have already noticed that, in order to "defend" the concept of singularity, has appeared a sort of sophistry or opinion like: one cannot really argue with a mathematical point. The opinion of this sort, however, cannot be really valid at all! Why? Clearly, when a mathematical point is used to describe the phenomena in physics, it has to obey the rules of physics. Thus, this sort of opinion, being merely a fabrication or product of covertly changing concepts or switching subjects, is utterly invalid, thereby totally futile in "defending" the concept of singularity. (Then what is a vivid example of changing concepts or switching subjects? Please allow me to show it. In a public lecture on the importance of good nutrition to our human beings, after showing various observational results, the lecturer concluded in the end: "Therefore, for a specific individual, the better nutrition one has, the stronger he or she is." A person in the audience disagreed immediately with: "The forage of horses is much less nutritious than the food of people, but horses are much stronger than people.") In addition, it should be cautioned that even the argument, which claims or views singularity as *a mathematical point*, still cannot be a truly valid excuse for "defending" the concept of singularity. Why? Mathematics never allows two rules that are clearly and seriously self-contradictory to coexist. For example, mathematics never permits 'two plus three equals five' and 'two plus three equals six' to exist together.

Interestingly or somewhat surprisingly, the information presented above can bring us something more important than this information itself. What seems to be ironically true is that, while the sort of opinion or sophistry discussed above, being represented with 'one cannot really argue with a mathematical point', does not and cannot change the simple *fact* that the concept of singularity is clearly and seriously self-contradictory, it does teach us a gravely bitter lesson (which can also be regarded either as a profound implication or as a constructive inspiration from the long-term viewpoint of science advancement). This can be specified as: in science, when a theory has incurred a clearly and seriously self-contradictory concept, a rational approach should carefully examine whether the theory is really correct or not (especially when *the* theory itself is widely known to be clearly and seriously inconsistent or contradictory with another extremely important theory), rather than keep looking for various excuses or "interpretations" for such a concept. There are no valid excuses or "interpretations" at all for any clearly and seriously self-contradictory concept! Any effort or action of looking for any excuse or "interpretation" for a clearly and seriously self-contradictory concept is no less than looking for a pretext for an assertion like a 2-kilogram hen produced a 5-kilogram egg. The history of science development has clearly told us: incorrect thinking is the number one enemy of science; correct thinking is the number one friend of science!

In addition to the first, also the most obvious or noticeable, feature that the concept of singularity turns out to be clearly and seriously self-contradictory, it is very difficult to imagine that singularity is a physically possible concept in fact—the second feature of singularity. Whenever people read the word *singularity*, what immediately jumps into their eyes are often the various expressions like: singularity is infinitesimally small and infinitely dense; singularity has zero radius and infinite density; singularity is an infinitesimal dot or has an infinitesimally small volume, and so on. (Some dear readers might have already noticed the routine procedure like the above.) If one sets aside a few minutes to think over these expressions profoundly and comprehensively, it seems rather rational that he or she could easily perceive or sense that it is actually very hard to think of a valid mechanism that could squeeze a massive star, whose mass can be hundreds of times the mass of the sun, into a zero-size point, singularity; not to mention that the theoretical basis of singularity, general relativity, turns out to be incorrect, as mentioned earlier. In other words, believe it or not (admit it or not), it is extremely difficult to find a reasonable explanation that singularity is a physically possible concept. Of course, it seems quite safe to conclude or believe that no rational scientists would claim there are an *infinite* number of mechanisms that can cause a highly massive star to become a point of being infinitely small and infinitely dense. (Commentator: this feature of singularity, along with the considerable help from its first feature that the concept of singularity turns out to be clearly and seriously self-contradictory, it seems not difficult that one can easily perceive or realize that the so-called singularity is almost certainly an impossible concept in physics.)

The third feature of singularity is that general relativity itself breaks down at singularity. In other words, the concept of singularity actually tells us: general relativity has created a place (singularity) where general relativity itself is no longer workable. This feature is thus a bit like the situation in which a driver encounters a dead end. A dead end means a driver has gone along wrong routes; the singularity in a black hole seems to indicate that the theory of general relativity may be incorrect—in fact, it turns out that general relativity cannot be correct, as shown earlier. (Related question and answer: what is the key to quickly grasping the three features of singularity discussed above? Answer: the theoretical basis of singularity, general relativity, does not have the ability to solve the most fundamental problem in front of itself, *why* space and time are variable thus relative in a gravitational field.) (Having been aware of the three features of singularity analyzed and discussed above, it seems to be a reasonable or realistic estimate that some, even many, insightful and sharp readers could clearly and easily realize that the concept of singularity turns out to be a big impediment that prevents one from really understanding the idea of the black holes described in *A Brief History of Time*. With this realization, it seems to be rather rational if these readers think of or ask the question like: can we find a new theory of black hole that shows there is no singularity at all within a black hole? Or can such a new theory reveal that the concept of singularity is actually unnecessary in explaining and describing black holes? This question is indeed very important. Dear readers will see such a new theory in the later part of this chapter, I promise. But at this moment, I hate to divert the attention of dear readers away from the central theme of this chapter—digging out the crux or root causes of why it's very difficult to understand the idea of the black holes presented in *A Brief History of Time*.)

Now, we can clearly see that there are four big barriers to one's understanding the idea of the black holes described in *A Brief History of Time*. First barrier: this idea cannot be correct, because its theoretical basis, general relativity, turns out unable to be correct. This barrier, on its own, is big enough to tell us: it turns out to be reasonable that one has had difficulty understanding this idea. Second barrier: this idea says that a black hole contains a singularity, whereas the concept of singularity turns out to be clearly and seriously self-contradictory. This barrier alone is sufficient to tell us: it turns out to be rational or realistic that one has encountered great difficulty in understanding this idea. So these two barriers jointly determine that: it turns out to be rational and fair that one could not really understand this idea. Third barrier: it seems, if not totally impossible, extremely difficult to find a valid explanation that singularity is a physically possible concept. As a result, these three barriers collectively determine that: it turns out to be reasonably understandable that one could not really understand this idea. Fourth barrier: at the singularity of a black hole, general relativity itself breaks down; not to mention that general relativity turns out to be incorrect. All in all, these four big barriers together have determined and demonstrated that: realistically and/or reasonably, it turns out to be quite understandable that one could not really understand the idea of the black holes described in *A Brief History of Time*. (Commentator: yes, with these four big barriers, it turns out to be both reasonable and fair that one could not really understand this idea. In fact, due to the solid existence of these four big barriers, it seems not only rational but also objective and fair to conclude that it would be really and truly weird if one claimed that he or she had really understood this idea.)

After finishing the check on the concept of singularity, now let us turn our eyes to the event horizon (the boundary of a black hole). According to the idea of the black holes described in *A Brief History of Time*, a black hole has two components: singularity and the event horizon, as mentioned above. Specifically, in his book *A Brief History of Time*, Hawking pointed out that: the event horizon of a black hole is formed by the light rays that just fail to escape from the black hole. So, while both components of a black hole are closely related to light, its event horizon is more directly connected with light; and light is always an indispensable, also crucially important, subject whenever black holes are talked about.

This prominent point can be clearly seen from the present definitions of black holes, like: a black hole is an object so massive that even *light* cannot escape from it; if enough mass is concentrated in a small enough region, the pull of gravity in such case becomes so strong that not even *light* can escape this region; and so forth. And more prominently, the event horizon appeared as a crucially important concept in the idea of the black holes described in *A Brief History of Time*. Accordingly, as an action of going the extra mile in checking whether this idea could really be understood or not, the third important thing to do is to inspect whether the event horizon is a reliable concept or not.

Before carrying out this inspection, we need to know about the origin of the event horizon. The event horizon originates from the *momentum* of photons (momentum is a quantity of motion of a moving object), being shown in the following known procedure leading to the birth of the event horizon. First, starting with the argument like: though a photon has no rest mass, it nevertheless behaves in collision as though it has an inertial mass, *because a photon has momentum*, the same feature as the inertial mass of an ordinary object, such as a car or a bicycle (the inertial mass of an ordinary object is defined based on Newton's second law. According to this law, an object with a certain amount of mass will accelerate, or change its speed, at a rate that is proportional to the magnitude or value of the net force acting on the object; this 'a certain amount of mass' is defined as the inertial mass of the object). The assumption of 'photons have inertial mass' (in general relativity) was, therefore, proposed by ascribing the momentum of photons to their assumed inertial mass. Then, by another assumption ('the equivalence of inertial and gravitational mass', also in general relativity), it was inferred that photons have gravitational mass thus being acted by gravity. Finally, the event horizon was derived from the escape velocity of an ordinary object that has a certain amount of inertial mass. (It should be noticed that, as the mathematical reflection of the procedure above, the event horizon of a black hole was usually derived and/or explained through the following three representative steps. First, determining the escape velocity of an ordinary object having a certain amount of inertial mass, with equation A that is from Newtonian gravitation theory. Then, directly by letting this escape velocity equal the speed of light in equation A, obtain equation B. Finally, by finding the solution of equation B, obtain the event horizon, which is expressed as the radius circling around a massive celestial body, or a black hole.)

Through the pieces of information presented above, one can clearly see that the event horizon was obtained along the route: the momentum of photons → photons have inertial mass → photons have gravitational mass → photons are acted by gravity, then by replacing the escape velocity of an ordinary object with the speed of light → the event horizon. This route shows that the *momentum* of photons is indeed the origin of the event horizon. Therefore, the task of inspecting whether the event horizon (the boundary of a black hole) is a reliable concept can be carried out through inspecting whether the current method of calculating the momentum of photons is right or wrong. That is, if this method is correct, we can say that the event horizon is a reliable concept; on the contrary, if this method turns out to be wrong, we cannot come to the conclusion that the concept of the event horizon is reliable. (Related question and answer: more specifically thus more directly, what is the relation between the concept of the event horizon and this method? Answer: the concept of the event horizon originates from the *momentum* of photons; this method determines the value of the *momentum* of photons.)

As a preparation to carry out this task, we need to know or review some basic knowledge about the momentum *change* when two objects collide (momentum is a quantity of motion of a moving object, measured as its mass multiplied by its velocity). Let us use the following situation as a simple and clear example. There are two objects, objects A and B, and immediately before their collision, *the momentum of object A is far greater than that of object B*—that is, compared to the momentum of object A, the momentum of object B is so small that it can be practically negligible. (You can think this situation as that, object A is like a racing vehicle in a sport competition running at a velocity about two hundred miles per hour, whereas object B is like a bicycle going at a velocity about ten miles per hour.) Let us check, with two different methods, the momentum of objects A and B immediately *after* their collision, in the following two cases.

First case: immediately after object A collides object B, method X calculates that the momentum of object B is more than *200 times* the original momentum of object A; method Z calculates that the momentum of object B is about 97 percent of the original momentum of object A. The related question is: even merely as a rough estimate, which method is wrong and which method is right? The answer is crystal clear: method X is definitely wrong because it incurs obviously ridiculous result; a bit like that a 2-kilogram hen produced a 50-kilogram egg! Method Z is probably correct because its result seems to be reasonable.

Second case: immediately after object A collides object B, and under the condition that the momentum of object A has not significantly reduced in the collision (such as the momentum of object A after the collision is still about 95 percent of its original momentum), method X calculates that the momentum of object B is about *two times* the original momentum of object A; method Z calculates that the momentum of object B is about 30 percent of the original momentum of object A. Again, the related question is: even just as a rough estimate, which method is wrong and which method is right? The answer is very clear: method X is undoubtedly wrong because it incurs clearly unreasonable result; method Z is probably correct because its result seems reasonable.

Now, let us return to our task: inspecting whether the current method of calculating the momentum of photons is correct or not. (This method was proposed by Einstein in 1916. According to this method, the momentum of a photon is equal to the energy of the photon divided by the speed of light; and the energy of the photon equals Planck constant, which is a fundamental constant in physics, multiplied by the frequency of the photon.)

One inspection is performed by an example about the photoelectric effect, being the phenomenon of ejecting electrons from a metal when it is shone by high-energy light. (The photoelectric effect tells us: when high-energy light shines on a metal surface, electrons are ejected from the metal surface. The phenomenon of the photoelectric effect was of great importance in the early 20th century for its experimentally demonstrating the particle-like nature of light, the feature simply called photons today, which are the tiny and discrete quanta or particles of light. It should be well aware that the photoelectric effect, viewed from the angle of photons, shows us such a clear fact: when a high-energy photon strikes an electron, the electron is ejected from metal surface; this fact is explicitly reflected in the explanation of the photoelectric effect.) This example is: ultraviolet light of wavelength 150 nanometers falls on a chromium electrode (*one thousand million (1 with nine zeros after it) nanometers = one meter, or one million nanometers = one millimeter; millimeter is usually the smallest graduations on a ruler or a meter stick). In this example, I have calculated that, with the current method of calculating the momentum of photons, the momentum of an ejected electron is more than *240 times* the original momentum of the photon that strikes the electron. Thus being quite similar to the first case above, such a result clearly shows that this method turns out to be definitely wrong, because it incurs obviously ridiculous result (by clearly violating momentum conservation law, one of the most fundamental principles in physics. About this law, please see the Glossary of this book if necessary); and again, a bit like that a 2-kilogram hen produced a 50-kilogram egg! (Related question and answer: where is the current method of calculating the momentum of photons wrong? Answer: this method greatly underestimates the momentum of photons, especially when the energy of a photon is very high—that is, the wavelength of a photon is very short, or its frequency is very high. This answer will be seen even more clearly, in the next paragraph, in comparison to the result calculated with another method.)

By contrast, in the *same* example, I have also calculated that the momentum of an ejected electron is 97 percent of the original momentum of the photon that strikes the electron, using electron-photon momentum relationship (which was discovered and verified just a few years ago by me). This relationship, being the combination of momentum conservation law (one of the most fundamental principles in physics) and the law of an orbiting electron with periodic impulses emitting photons (the main content of this law will be briefly introduced in chapter six), reveals the connection in momentum between an electron and its emitting photons *for the first time in the history of science*. This relationship shows that: the momentum of a photon comes from and is equal to the momentum change of the electron emitting the photon over this electron's one complete impulse period (in both value and direction); and the momentum of a photon is equal to the orbiting momentum of the electron emitting the photon (in value). It should be noticed that, though electron-photon momentum relationship gives the reasonable result, the purpose of presenting this result is to display, through a sharp comparison, that the current method of calculating the momentum of photons turns out to be clearly wrong. And this purpose should be the focus of dear readers here.

Another inspection is carried out via an example, in which a high-energy photon, such as an X-ray photon, strikes an electron that is almost stationary, which means that the momentum of the electron is much, much smaller than that of the X-ray photon striking it (that is, compared to the momentum of the X-ray photon, the momentum of the electron is small enough to be practically or safely negligible). (Such an example has realistic and factual grounds. For instance, in the experiment of the famous Compton scattering, conducted in 1922, X-rays were used to strike electrons. The Compton scattering was widely recognized as important, independent evidence supportive of the particle-like behavior of light.) In the example, I have calculated that, using the current method of calculating the momentum of photons, the momentum of an electron (after being struck by an X-ray photon) is about *two times* the original momentum of the X-ray photon that strikes the electron, whereas the X-ray photon loses only about 5 percent of its momentum in the striking (that is, the X-ray photon still keeps about 95 percent of its original momentum after the striking). So, being very similar to the second case above, this result clearly tells us that the current method of calculating the momentum of photons cannot be correct, because it leads to obviously unreasonable result. (For comparison, in the *same* example, I have also calculated that, with electron-photon momentum relationship just mentioned above, the momentum of a struck electron is about 30 percent of the original momentum of the X-ray photon that strikes the electron.)

The explicit conclusions from these two inspections have thus uncovered or shown the irrefutable fact that the current method of calculating the momentum of photons turns out to be clearly wrong, because it has led to obviously ridiculous, plainly unreasonable results. This uncovered fact has no choice but to tell us such a clear, unavoidable *reality:* the event horizon (the boundary of a black hole) turns out to be an unreliable concept, because it originates from the momentum of photons, as analyzed and shown above; and because it is this very method that determines the value of the momentum of photons, as pointed out above. (Having realized that the concept of the event horizon turns out to be unreliable, it seems neither irrational nor inappropriate if some, even many, perceptive and curious readers think of or raise the related question like: can we find a new theory of black hole that redefines the boundary or size of a black hole? Dear readers will see such a new theory after a few pages in this chapter. But right now, please don't divert your attention away from the central theme of this chapter—finding out the crux or root causes of why it's not easy to understand the idea of the black holes described in *A Brief History of Time*.)

Related and deep discussions (this matter can be optional, though some readers may realize that this kind of discussions turns out to be not only very interesting and inspiring, but also quite meaningful and helpful). Having known the irrefutable *fact* that the current method of calculating the momentum of photons turns out to be clearly wrong, it seems neither unusual nor unreasonable that some readers may have the possible or potential reactions like: surprised, even shocked, hard to believe, or regretted. Each of these reactions is understandable or quite normal, considering this *fact* is probably too new or too radical in their eyes. Nevertheless, is there something more important than these reactions themselves? Or more specifically, what bitter lesson can we draw from this irrefutable *fact* or such a grave mistake? To this question, different people may have different answers. Though I firmly believe that some, even many, readers are well able to have better answers than me, I would still like to share mine with dear readers here (certainly, all readers are welcome to have comments on my answer). My answer is: in science, it is crucially important to examine an idea or method from the different situations or processes that are covered by the idea or method, provided that these situations or processes are inherently related with each other, in order to ensure that this idea or method can give an overall coherent picture. In other words, it is fundamentally necessary to advocate and encourage the mode of comprehensively and systematically thinking in science, in order to avoid the ridiculous or obvious mistakes like in the fable of 'the Blind Men and the Elephant'—this fable has been mentioned in chapter two.

This answer can be specifically reflected in the two historical events or instances. First instance: Einstein proposed the current method of calculating the momentum of photons in 1916, and he also interpreted the photoelectric effect in 1905. So if he had gone back and checked this method with the related result in the photoelectric effect, the *fact* that this method turns out to be ridiculously wrong could have been identified in the 1910s. Second instance: the famous experiment of the Compton scattering was conducted in 1922, so, and again, if Arthur Compton had checked this method with the related result in his experiment, the *fact* that this method turns out to be clearly wrong could have been detected in the early 1920s. Yes, while there is an opinion saying that history has no assumptions or 'IFs', many people might have already noticed that history, including the history of science of course, does have some amazing similarities. And some of these similarities have brought us some deep regrets. For example, another famous regret is that Aristotle's idea, which said that a heavy body should fall faster than a light one, had dominated the world for nearly two thousand years, no one until Galileo Galilei showed this idea turned out to be wrong. In a sense, we should not blame former generations for these regrets, instead what seems to be really or more important is that we should draw bitter lessons from these regrets. Thus, in this sense, we should not regret too much about the grave mistake in the current method of calculating the momentum of photons—this method only has a history of just one century.

After the discussions above, let us still return to the concept of the event horizon. Let us go to another feature of this concept, its self-contradictory feature, because such a feature can considerably help us further think over the reality that the event horizon (the boundary of a black hole) turns out to be an unreliable concept. (Commentator: of course, through this self-contradictory feature, one can see this reality more clearly, thus accept it easily and quickly.) Then what is the behavior or nature of this self-contradictory feature?

As mentioned above, in deriving the event horizon, the escape velocity of an ordinary object with a certain amount of mass was forcibly replaced with the speed of light by making the former equal the latter. However, as pointed out in chapter two, special relativity says that the speed of light is an unattainable speed for *any* ordinary object that has a mass, and *any* objects with mass must move at the speed less than that of light, no matter how to increase their speed. In other words, special relativity does not allow the velocity of an ordinary object having a mass to jump to the speed of light; that is, special relativity does not permit the appearance of the concept of the event horizon at all! On the other hand, this concept was developed crucially based on general relativity, which is clearly reflected in the fact that this concept is indispensably dependent on the two assumptions of general relativity (one is 'photons have inertial mass'; another is 'the equivalence of inertial and gravitational mass'), as mentioned above. That is to say, general relativity does allow the appearance of this concept. Therefore, it becomes clear enough that the concept of the event horizon turns out to be plainly self-contradictory, when viewed from the angles of special relativity and general relativity. Undoubtedly, such a self-contradictory feature can substantially help one further realize the reality that the event horizon turns out to be an unreliable concept. (Note: though in chapters two and three, we have clearly seen that special relativity and general relativity cannot be correct, dear readers should be aware that the focus here is on the self-contradictory feature of the concept of the event horizon.)

Up to now, one can clearly see that there are three major reasons showing us why it's difficult to understand the idea of the black holes described by Hawking in his book *A Brief History of Time*. First reason: it turns out that this idea cannot be correct, because its theoretical basis, general relativity, turns out to be incorrect. This reason alone is sufficient to tell us: it turns out to be rational or normal that one has had great difficulty following this idea. Second reason: this idea says that a black hole contains a singularity, whereas the concept of singularity turns out to be clearly and seriously self-contradictory. This reason itself is enough to tell us: it turns out to be rational thus understandable that one could not really understand this idea. So these two reasons together have determined and demonstrated: it turns out to be rather rational thus pretty understandable that one could not really understand this idea. Third reason: this idea says that a black hole has the event horizon (the boundary of a black hole, another important component of black holes); nevertheless, it turns out that the event horizon cannot be a reliable concept. Therefore, these three reasons—actually they are three hard facts or three pieces of hard evidence, have collectively and clearly shown us such a simple and clear fact: there are more than enough reasons to justify that this idea could not be really understood. (Commentator: moreover, it seems quite reasonable or realistic that some, even many, insightful and sharp readers might have been well aware that the three reasons above, or the three hard facts above, are actually three impassable great barriers preventing one from really understanding the idea of the black holes described in *A Brief History of Time*. This awareness can naturally or easily enable these readers to conclude: it turns out to be neither rational nor valid if one holds that he or she has really understood this idea. Such a conclusion seems to be a rationally acceptable consensus to the people who have fully realized these three reasons; this rational consensus can be a great relief or wonderful relaxation for the people who have been struggled or perplexed for having not really understood this idea.)

More specifically thus more directly, these three reasons, being three hard facts or three pieces of hard evidence, clearly show and explicitly manifest the crux or essence of why it's difficult to understand the idea of the black holes described in *A Brief History of Time*. This is because these three pieces of hard evidence fully reveal this crux from three different directions or angles. The first direction: the theoretical basis of this idea, general relativity, turns out unable to be correct. The second direction: one component of a black hole according to this idea, the so-called singularity, turns out to be a concept that is clearly and seriously self-contradictory. The third direction: the other component of a black hole according to this idea, the event horizon, turns out to be an unreliable concept. Therefore, this crux actually focuses on such a newly revealed or clearly shown fact: it turns out that this idea itself cannot be correct, or this idea turns out to be incorrect. With and through this crux, one can naturally, or rationally, be aware of the secrets of why it's difficult to understand this idea: in science, the very basic feature of an idea that cannot be correct is that *the* idea cannot be really understood in truth, because it is clear, even obvious, also self-evident, that rational people with reasonable thinking could not really understand an idea that turns out unable to be correct. (Commentator: yes, that is clearly true and surely valid. More than that, if anybody claimed that he or she could really understand an idea that turns out to be incorrect, other people might naturally question whether this person has a correct or rational thinking. In addition, it seems quite reasonable or realistic that some, even many, sensible or insightful readers might have such a rational attitude: even if some people claimed that they could understand an idea that turns out unable to be correct, the so-called "understanding" of this sort is actually meaningless or pointless in science. This rational attitude seems quite obvious to most people, even if not all people.)

Having realized the crux of why it's difficult to understand the idea of the black holes described in *A Brief History of Time* turns out to be that this idea cannot be correct, it seems neither unreasonable nor illogical that this realization may bring about two types of rational or possible, also quite insightful, reactions from dear readers. One is reflected in the question like: are there other severe consequences that come from this crux? (Or does this crux have other profound implications?) Another is embodied in the question like: are there available clues or signals that can substantially help us chew over and digest this crux? (Or are there available clues or signals that can enable us to see this crux more clearly?)

The answer to both types of the questions above is YES. One of the disastrous consequences is that this idea is not only unable to resolve the four long-standing, fundamentally important problems in physics (they are: the mystery of dark matter, the mysterious source of gamma ray bursts and ultrahigh-energy cosmic rays, as well as the tricky puzzle of the GZK paradox—about this famous paradox, please see the Glossary of this book if necessary. Worldwide many scientists have been working on these four big problems for a long time, yet still fruitless—not even a hint of real success in fact), but also becomes the shackles and obstacles to solving them. What should be pointed out is: as long as this idea stands there as an impassable and unavoidable great obstacle, it is definitely not possible to solve these four big problems—no matter how much data has been and will be collected; no matter how much money and effort have been and will be spent. What should be noticed or clarified is: if this idea had been correct, these four big problems would have already been solved! In other words, believe it or not (admit it or not), these four big problems have actually become the clear clues that can substantially help one further perceive and realize the newly revealed or clearly shown fact that this idea cannot be correct. (Commentator: thus, if some people, especially some of those respected professionals in physics, still have some psychological obstacles in facing the newly revealed fact that the idea of the black holes described in *A Brief History of Time* turns out to be incorrect, please think of, think about, and deeply ponder these four big problems. They are really quite helpful! Once one has realized this newly revealed fact, he or she can naturally, or easily, be aware of the reality that this idea cannot be really understood, because it is quite normal that rational people with reasonable thinking could not really understand an idea that turns out to be incorrect.)

Before ending this chapter, I shall fulfill my promise made earlier: I shall introduce a new theory of black hole that shows us there is no singularity within a black hole, and that redefines the boundary or size of a black hole. In fact, besides my promise, there are far more important reasons that such a new theory should be introduced. But only a few main reasons will be mentioned here.

First, this new theory provides a sharp contrast through which one can see more clearly the newly revealed or shown fact that the idea of the black holes described in *A Brief History of Time* turns out to be incorrect. Many of us might have such an experience: in order to have a clearer and better understanding of something (an idea, a theory, or the conclusion to an idea or a theory, for example), we often need a contrast or comparison, and sometimes even want a sharp contrast or tangible comparison. So is in our distinctly perceiving and explicitly understanding this newly revealed fact. (Narrator: having a good understanding of this newly revealed fact is one of the most effective ways or methods to grasp the crux of why it's difficult to comprehend the idea of the black holes said in *A Brief History of Time*, because in science, it is self-evident and quite obvious that rational people with reasonable thinking could not really understand an idea that turns out to be incorrect. And having a good grasp of this crux is the most important purpose of this chapter, also the take home message of this chapter.)

Second, this new theory gives an easily comprehensible explanation to the observed phenomenon of black holes. If there were no such a new theory, even having recognized the reality that the idea of the black holes (said in *A Brief History of Time*) cannot be really understood, it would still be understandable that some people, especially some of the respected professionals in physics or astronomy, might have the reaction like: yes, it is indeed very difficult to imagine that one could really understand this idea—either in essence or in fact, but it is still much better than having no idea on how to deal with the important phenomenon of black holes. In addition, I cannot rule out the possibility that some, even many, insightful and sharp readers may have such an expectation: after knowing about a new theory that is easy to understand, one can see more clearly, through a striking contrast, the reality that the idea of the black holes described in *A Brief History of Time* cannot be really understood in truth. So with these important reasons, it seems not inappropriate to introduce a new theory of black hole before closing this chapter. (Of course, dear readers will have a final say on whether it is really appropriate to introduce such a new theory after seeing it.)

Let us start from previewing the two basic features of the new black hole theory (which was developed just a few years ago by me). First, as its fundamentally noticeable mark, this new theory is of mechanism-revealed nature, being clearly determined by two sufficient reasons. One is that its theoretical foundation is mechanism-revealed. This new theory is based on the newly discovered and verified, mechanism-revealed gravitational theory that reveals the mechanism of *why* space and time are variable thus relative in a gravitational field (some of its main content has been concisely introduced in the later part of chapter three). Another is that this new theory reveals the mechanism behind its describing phenomena. For these two reasons, this new black hole theory has been named as *mechanism-revealed black hole theory* (MRBHT, for short) by me. Accordingly, black holes, when they are explained with MRBHT, are referred to as *mechanism-revealed black holes*. What should be pointed out or noticed is: the adjective word *mechanism-revealed* is to distinguish MRBHT from all other black hole theories that are based on general relativity. (Related reminder: as mentioned in chapter one, general relativity is based on at least five assumptions, hypotheses and postulates, so general relativity is a *postulate-based* theory; thus all the black hole theories based on general relativity are ultimately *postulate-based*. Accordingly, black holes, when they are interpreted with the black hole theories based on general relativity, are actually *postulate-based black holes*.) (Narrator: this fundamental feature, therefore, shows that MRBHT is essentially and contrastingly different from the idea about the black holes described in *A Brief History of Time*, because this idea comes from the black hole theories based on general relativity, as mentioned at the beginning of this chapter. Such a crucial, also marked, difference seems to be a good reminder of the great importance and/or necessity to introduce MRBHT.)

Second, the characteristic, also noticeable and impressive, feature of this new black hole theory (that is, MRBHT stands for mechanism-revealed black hole theory) lies with: it is easy to understand. For instance, the theoretical core of MRBHT is similar in essence to the following plain principle. Let us say, you measure the distance between two locations as 760 miles on one map, whose scale is 1:250. If you measure this distance on another map whose scale is 1:500, but *still using the scale of 1:250*, you will measure this distance as 380 miles. Why is MRBHT so simple? The fundamental reason is that MRBHT is of mechanism-revealed nature, as just pointed out above; the direct reason is that MRBHT is mathematically pretty simple: the mathematics at the level of high school or first-year undergraduate is enough.

Having previewed the two basic features of this new black hole theory, we are about to enter its mechanism and essence—the specific and key information of this theory. As mentioned earlier in this chapter, whenever a black hole is talked about, light is always an indispensable, also crucially important, subject. So, as a preparation to enter this new theory, let us review the basic concept about visible and invisible light first.

This concept includes two aspects. First aspect, whether a light is visible or invisible depends on its wavelength or frequency (the wavelength of a light is the distance between its one wave crest and the next; the frequency of a light is the number of its waves per second). For example, it has been known that the wavelength of visible light is in the range from 380 to 760 (or 390 to 780) nanometers, with violet light between 380 to 440 nanometers, and red light between 620 to 760 nanometers ([*]one thousand million nanometers = one meter, or one million nanometers = one millimeter; millimeter is usually the smallest graduations on a ruler or a meter stick). Second aspect, the wavelength of a light is determined by the length scale used to measure its wavelength (correspondingly, the frequency of a light is determined by the time scale used to measure its period or frequency; its period times its frequency is equal to one). For instance, the wavelength of a light is measured as 700 nanometers (that is, a red light) by observer A at location A where the local length scale is 1.00. When the same light travels to location B where the local length scale is 0.60, what is the wavelength of the same light with respect to observer A (that is, still with respect to his 1.00 length scale)? Answer: the wavelength of the same light is equal to 420 nanometers, which is from 700 nanometers times (0.60/1.00), thus becoming a violet light.

Now, let us go to the *mechanism* and *essence* of this new black hole theory (that is, mechanism-revealed black hole theory). According to the newly discovered and verified, mechanism-revealed gravitational theory (some of its main and key results have been briefly introduced in the later part of chapter three), the length scale and time scale in the gravitational field of a massive celestial body increase radially outward (decrease radially inward) (Fig. 4.1). That is, the farther away from the body, the larger the scales of length and time; the closer to the body, the smaller the scales of length and time. As a result, when the scales of length and time nearby a hugely massive celestial body are reduced to such an extent that all visible light becomes invisible, a black hole is thereby formed. For example, the wavelength of a light is measured as 700 nanometers (that is, a red light) by observer George at a location very far away from a hugely massive body, where the length scale is infinitely close to 1.00 (thus can be practically regarded as 1.00). When the same light travels to a location very near to the hugely massive body, where the length scale is 0.50, to observer George (of course, still with respect to his 1.00 length scale), the wavelength of the same light will become 350 nanometers (thus becoming an invisible light)—this 350 nanometers comes from: 700 nanometers times (0.50/1.00).

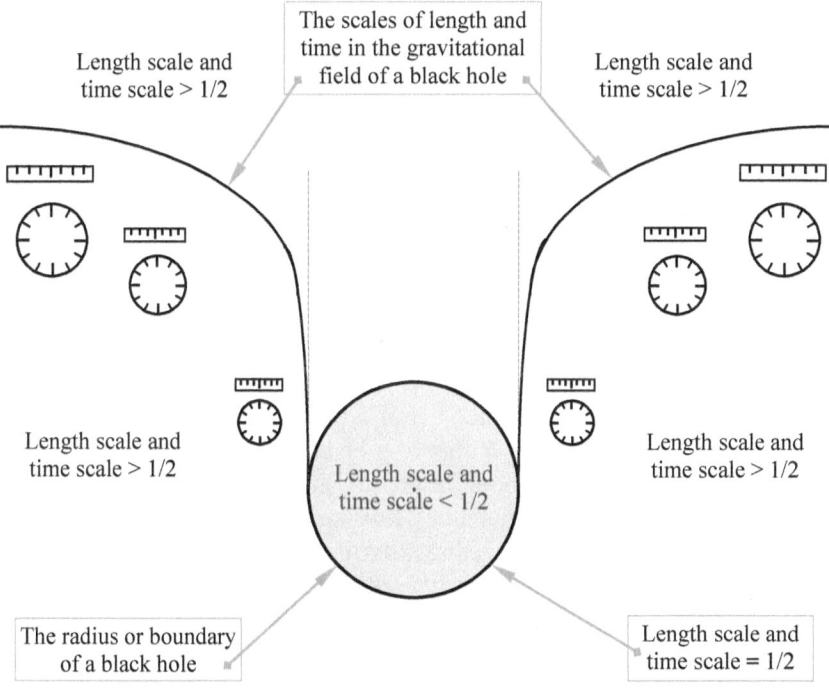

Figure 4.1, mechanism-revealed black hole theory (a new black hole theory). This new theory reveals the mechanism of black holes, and shows the essence of black holes, thus solves the most fundamental, the most essential and the most crucial problems about black holes.

So a black hole is a region where all visible light becomes invisible: all the visible light entering the black hole becomes invisible; all the light emitted from the black hole is also invisible in the region of the black hole. Resultantly, to very distant observers, there are no visible rays of light at all in the region of the black hole—this region looks black. Therefore, the *mechanism* of mechanism-revealed black hole theory is that the existence of a hugely massive celestial body reduces the scales of length and time in its vicinity to such an extent that all visible light becomes invisible; thus the *essence* of a black hole is the tremendous reduction in the scales of length and time that occurs in the gravitational field of a hugely massive celestial body. In addition, what should be noticed is: the length scale and time scale in the enormously large extending region that radially spreads out from a black hole are significantly reduced too (due to the existence of the black hole), though not as intense and noticeable as within the region of the black hole.

After knowing about the mechanism and essence of mechanism-revealed black hole theory (MRBHT), let us see the size of a black hole, determined with MRBHT. In order to calculate the size of a black hole, we need to know the meaning of the ratio of visible light boundary wavelength (assigned as W_{vlb}). The idea of W_{vlb} is from color index, the ratio of the wavelength of violet to red light. As mentioned above, the wavelength of visible light is in the range from 380 to 760 (or 390 to 780) nanometers ([*]one thousand million nanometers = one meter), which means that the lower and upper boundaries of visible light's wavelength are respectively 380 and 760 nanometers (or 390 and 780 nanometers), with violet light close to the lower boundary, and red light close to the upper boundary. Thus W_{vlb} is equal to 380/760 (or 390/780) = 1/2. So, W_{vlb} is the result of an application of color index, by extending color index to the very edge of the wavelength of visible light, in order to reflect the black feature of black holes.

Now, we are ready to see the size of a black hole, represented with its radius (assigned as R_{bh}). The radius of a black hole is the threshold radius that marks the black hole's boundary, on which entering visible light just begins to become invisible, whereas leaving light just begins to become visible. This requirement determines that the radius of a black hole must combine the mechanism of mechanism-revealed black hole theory with the meaning of W_{vlb} (this meaning has been clearly elucidated in the paragraph above), in order to provide a reasonable thus easily comprehensible explanation for the observed phenomenon of black holes.

This combination determines that the value of radius R_{bh} corresponds to the value of r in the quantity $GM/(rc^2)$ that makes: length scale = W_{vlb} = 1/2 (where M is the mass of a hugely massive celestial body that results in a black hole, r is the radius from the mass center of the mass M, G is the universal gravitational constant, and c is the speed of light. So the quantity $GM/(rc^2)$ is unitless. *Oh, by the way, if the r in the quantity $GM/(rc^2)$ is equal to R, where R is the radius of a celestial body, this quantity thus becomes $GM/(Rc^2)$. It should be mentioned that $GM/(Rc^2)$ is one of the most important quantities in modern astronomy; it is no exaggeration to say that all today's astronomers are very familiar with this specific quantity. That is, such an important quantity in astronomy, $GM/(Rc^2)$, turns out to be a special case of the quantity $GM/(rc^2)$; this feature can enable one to realize the meaning and importance of the quantity $GM/(rc^2)$ more clearly*). $GM/(rc^2)$ is a crucial term in the core equation of mechanism-revealed gravitational theory (this theory has been briefly introduced in the later part of chapter three; and its core equation, which is the equation of gravitational scales of space and time, determines the length scale and time scale in the gravitational field of a massive celestial body). Then, by letting the length scale in this core equation be equal to $W_{vlb} = 1/2$, obtain $GM/(rc^2) = 2$. Finally, from $GM/(rc^2) = 2$, obtain $r = R_{bh} = GM/(2c^2)$, where M is the mass of a black hole, G and c have the same meaning as above. (For comparison, the value of this radius is exactly one-fourth of that of the well-known Schwarzschild radius.)

Now, one can see that according to mechanism-revealed black hole theory, the size of a black hole, when expressed by its radius, is proportional to its mass; to be exact, the size of a black hole entirely and only depends on its mass, or the size of a black hole is totally determined by its mass. (Interestingly, this basic feature is clearly similar to the conclusion that the size of a black hole depends only on its mass; this conclusion has appeared in *A Brief History of Time*. And this basic feature is also quite similar to the result that the area of a black hole increases whenever matter falls into it; this result has also been mentioned in *A Brief History of Time*. Quite obviously, these similarities can substantially enhance the acceptability of this new theory, at least to a certain degree or from a certain angle.) But what should be pointed out is: the radius of a black hole in mechanism-revealed black hole theory is fundamentally different from the event horizon of a black hole (the boundary of a black hole) in the idea of the black holes described in *A Brief History of Time*. The radius of a black hole is the dividing line between visible and invisible light, whereas the event horizon of a black hole is defined as the boundary from which light rays just fail to escape from the black hole; and moreover, the event horizon of a black hole turns out to be an unreliable concept, as analyzed and revealed above.

Mechanism-revealed black hole theory (MRBHT) reveals several fundamentally important natures of mechanism-revealed black holes. First of all and most of all, a mechanism-revealed black hole does not have the so-called singularity; in fact, a mechanism-revealed black hole does not need the concept of singularity at all. In other words, at any location or place within or around a mechanism-revealed black hole, there is utterly no self-contradictory point. (In contrast, the idea of the black holes described in *A Brief History of Time* says that there is a singularity inside a black hole. However, the concept of singularity turns out to be clearly and seriously self-contradictory, as analyzed and concluded above; not to mention that the theoretical basis of singularity, general relativity, turns out to be incorrect.) As a result—as a clear and noticeable result, this fundamental nature of MRBHT shows the fundamental and profound difference between MRBHT and this idea. Therefore, this fundamental and profound difference actually provides a sharp and noticeable contrast through which one can see more clearly the newly revealed or shown fact that this idea turns out unable to be correct, which can explicitly help one realize this newly revealed fact even more clearly and definitely. Once one has realized this newly revealed fact, he or she will have no difficulty grasping the crux of why it's difficult to understand this idea, because in science it is self-evident and obvious that rational people with reasonable thinking could not really understand an idea that turns out to be incorrect. Besides, this fundamental and profound difference is also quite helpful for readers to ponder the unavoidable reality that the idea of the black holes said in *A Brief History of Time* cannot be really understood even further, via the fact that this idea says that a black hole contains a singularity; via the fact that the concept of singularity turns out to be clearly and seriously self-contradictory; and via the plain and self-evident, also obvious and irrefutable, basic rationale in science: no one could really understand a concept that turns out to be clearly and seriously self-contradictory, thus no one could really comprehend an idea that contains such a concept.

Second nature: a mechanism-revealed black hole can emit light (photons) from the region within its radius. This nature enables mechanism-revealed black hole theory to be the key to solving the four long-standing, fundamentally important problems in physics: dark matter, gamma ray bursts, ultrahigh-energy cosmic rays, and the GZK paradox (this paradox has been concisely explained in the Glossary of this book). (On the contrary, according to the idea of the black holes described in *A Brief History of Time*, a black hole cannot emit light from the region within its boundary, the so-called event horizon. Consequently, this idea is not only unable to solve these four big problems, but has also become the shackles and obstacles to solving them.) What should be emphasized or clarified is that, though Hawking in his *A Brief History of Time* talked about the emission or radiation from black holes—that is, a black hole can emit X rays and gamma rays, this emission is merely from the place just outside the event horizon of the black hole, rather than from within the black hole. Please further notice that, the existence of such a sort of emission is under the fundamental *prerequisite* that nothing can escape from within the event horizon of a black hole. Moreover, this sort of emission is far too small to explain the tremendously huge source of gamma ray bursts (otherwise, the big puzzle of gamma ray bursts would have already been solved much earlier!). Therefore, there is a big and essential difference between the emission mentioned by Hawking and light (photons) emitted from mechanism-revealed black holes. (Narrator: quite obviously, this crucial difference is quite able to serve as a striking comparison or a clear clue that can considerably help one be well aware of the newly revealed or clearly shown fact that the idea of the black holes described in *A Brief History of Time* turns out to be incorrect. This awareness is one of the most effective ways for one to grasp the crux of why it's hard to understand this idea.)

In addition, the constituents of mechanism-revealed black holes are not essentially mysterious at all, in comparison with other ordinary celestial bodies. Please notice that, as revealed by mechanism-revealed black hole theory (MRBHT), the decisive, also essential, difference between black holes and other ordinary celestial bodies (such as a variety of stars and planets) lies with the hugely massive feature of black holes, rather than in their constituents. In other words, a (mechanism-revealed) black hole can be formed as a result of many ordinary celestial bodies joining together. For instance, according to this new black hole theory (that is, MRBHT), as long as the mass of a celestial body is greater or equal to several dozens of times the mass of the sun, the celestial body is or becomes a (mechanism-revealed) black hole, if the average density of the black hole is set at the density of an atomic nucleus. The available knowledge about the mass of various celestial bodies shows that such a requirement on mass is not difficult to meet even in our Milky Way galaxy, needless to say in the vast universe. (In contrast, according to the idea of the black holes said in *A Brief History of Time*, the constituents of black holes have been a great mystery; in fact, within the paradigm or stereotype of this idea, the constituents of black holes have actually become one of the long-term unsolved, fundamentally important problems in physics or astrophysics. Note: astrophysics is a branch of astronomy that deals with the physical and chemical structure of the stars, planets, etc.)

With and through these fundamentally important natures, one can clearly see that mechanism-revealed black holes are essentially and noticeably different from the black holes described in *A Brief History of Time*. With these essential differences as a substantial help, and with these noticeable differences as a sharp contrast or vivid comparison, it seems not difficult that one can clearly and distinctly realize the newly revealed or shown fact that the idea of the black holes described in *A Brief History of Time* turns out unable to be correct. Such a realization is the best way, or one of the best ways, of knowing the crux of why it's difficult to understand this idea, because it is pretty reasonable and fair that sensible or intelligent people with rational thinking could not really understand an idea that turns out to be incorrect. (Related questions and answers: concisely and ultimately, what is the root cause that gives birth to these essential and noticeable differences? Answer: the theoretical basis of mechanism-revealed black hole theory is the newly discovered and verified, mechanism-revealed gravitational theory that shows *why* space and time are variable thus relative in a gravitational field by revealing the mechanism behind this *why*, whereas the theoretical basis of the idea of the black holes said in *A Brief History of Time* is general relativity that is unable to solve the most fundamental problem in front of itself—*why* space and time are variable thus relative in a gravitational field. Questions: then could one clearly perceive and explicitly realize, even merely from the perspective of general relativity itself, the *fact* that general relativity is really unable to solve this most fundamental problem? How? Answer: yes; definitely and surely! The following simple and clear method is workable and quite effective. If general relativity had been able to solve this most fundamental problem, its postulate of 'invariant scales of length and time', which says that the scales of length and time at different points over an entire gravitational field are the same, would not have been necessary at all. And so, the irrefutable or undeniable *actuality* that general relativity indispensably and desperately necessitates this very postulate can enable one to perceive this *fact* clearly and realize it explicitly; of course, this irrefutable or undeniable *actuality* also makes one unable to deny this *fact*.)

All in all, this chapter has revealed, with more than sufficient hard evidence—being represented by the three main reasons summed up earlier (these reasons manifest even further, thus could be seen much more clearly via the striking contrasts against a new theory of black hole), the secrets of why it's difficult to understand the idea of the black holes described in the book *A Brief History of Time*. Having known these secrets, one can clearly and surely, though perhaps somewhat amazingly or unexpectedly, realize: the responsibility in which people could not really understand this idea turns out not to be on the people who have read *that book*. Accordingly, one shouldn't have felt discouraged or disappointed for having not really understood this idea. Such a realization is the ultimate goal of this chapter, also being a great consolation to me or a generous reward for me from dear readers.

CHAPTER 5

WHY DIFFICULT TO UNDERSTAND "THE ORIGIN AND FATE OF THE UNIVERSE"

[The Window on This Chapter]

One "poof", the universe had been abruptly created—the origin of the universe (interpreted in *A Brief History of Time*), according to general relativity.

Another "poof", the universe will suddenly disappear—the fate of the universe (described in *A Brief History of Time*), from the standpoint of quantum mechanics.

However, we have also been told such a <u>fact</u> in *A Brief History of Time* (1998 version, P. 12): general relativity and quantum mechanics cannot both be correct because they are known to be inconsistent with each other. Which "poof" do you believe then?!

Having revealed the secrets of why it's difficult to understand the expanding universe, are we still far from digging out the root causes of why it's hard to understand the origin and fate of the universe presented in *A Brief History of Time*? Are there some impassable barriers that hinder one from really understanding this origin and fate?

After reading this chapter, you will find out: it turns out to be rather rational thus quite understandable that/if one could not really understand the origin and fate of the universe interpreted and described in *A Brief History of Time*.

Overall, the ultimate or direct purpose of this chapter is to look into or dig out the secrets of *why* it's difficult to understand the origin and fate of the universe described in *A Brief History of Time* (in chapter eight, "The Origin and Fate of the Universe", 1998 version). The achievement of this purpose can enable one to be well aware: no wonder many people have run into great difficulties in understanding this origin and fate; it turns out to be quite reasonable and fair that one could not really understand this origin and fate. In other words, the actual effect of achieving this purpose will be a wonderful relaxation or great consolation for the people who have been perplexed or struggled for having not really understood this origin and fate.

In order to achieve this purpose, the first thing we need to do is to be familiar with the crucial and decisive, thus the most important feature of this origin and fate, because such a feature will explicitly tell us how to attain this purpose by specifying what actions we ought to take. Thus, it is this feature that will determine both the general direction and the specific content of this entire chapter. For that reason, let us start the journey of this chapter from knowing about this most important feature.

What is this feature then? Let us see it. As clearly reflected in the related content of *A Brief History of Time* (in chapter eight, "The Origin and Fate of the Universe", 1998 version), the view or interpretation about the origin and fate of the universe in *that book* is attached onto the "hot big bang model", because it is this model that describes the universe—this model gives a mathematical description of the expanding universe from about one second after the suggested big bang. In fact, this model is not only the key framework that forms this view, but also the heart and soul of this view, for two plain reasons. One is that the main theme of this model is about the universe, or about 'the expanding universe', to be exact; another is that this model is the most comprehensive, the most important, also the most influential, contemporary description of 'the expanding universe'. On the other hand, it should be noticed that all the other models or proposals mentioned in *A Brief History of Time*, by dealing with the universe *within* only about one second shortly after the proposed big bang, were merely for attempting to connect with this model. (This is a bit like that, if the hot big bang model is regarded as an entire large building, including its foundation of course, then all these other models or proposals are merely about some trivial details of the bottom line of the building's foundation, such as its shape and whether it is smooth or grooved.)

What does this feature tell us? It shows that: only if one could really understand the hot big bang model, could he or she truly understand the origin and fate of the universe explained and described in *A Brief History of Time*. It determines that: the *prerequisite* of truly understanding this origin and fate lies with really understanding this model. It warns or reminds that: if one could not really understand this model, it would be impossible to imagine that he or she could truly understand this origin and fate in fact; it would thus become meaningless or futile to talk about the issue of whether or not he/she could truly understand this origin and fate in effect. Therefore, this feature clearly tells us: in order to achieve the ulti-

WHY DIFFICULT TO UNDERSTAND "THE ORIGIN AND FATE OF THE UNIVERSE"

mate purpose of this chapter, we have to take on the inescapable task of inspecting whether one could really understand this model or not (this task has thereby become the central task of this chapter).

How to inspect whether or not one could really understand this model then? Of course, we need to check whether one could really understand what this model describes, the idea of the expanding universe, simply because this idea is the main theme of this model, as just mentioned above. (This is just like that, to inspect whether one could really understand a book is to check whether or not he or she could truly understand what the book says.) In other words, only if one could really understand this idea, could he or she really understand the hot big bang model, the least requirement that he/she could truly understand the origin and fate of the universe presented in *A Brief History of Time*. (That is, the very *prerequisite* that one can truly understand this origin and fate lies with that he or she can really understand this idea.) Accordingly, the central task of this chapter has turned into looking for the answer to the crucially, also decisively, important question: could the idea of the expanding universe be really understood in truth? Or could one really understand this idea in fact?

This central task reminds us of chapter three (its title is: Why Difficult to Understand "The Expanding Universe"), because the answer to the question above has been found there. Specifically, in that chapter, we have clearly seen that there are three big and impassable obstacles preventing one from really understanding the idea of the expanding universe. What are these obstacles? Let us review them as follows.

The number one obstacle is the revealed plain fact that the theoretical basis of this idea, general relativity, turns out unable to be correct, being sufficiently determined by either of the following two inescapable realities. (For the detailed or more information about these two realities, please go back to chapter three; both realities have also been clearly mentioned in chapter four. So only the key points about these two realities are briefly reviewed here.)

One reality is such an inability: general relativity is unable to solve the most fundamental problem in front of itself, *why* space and time are variable thus relative in a gravitational field. For instance, while general relativity tells us that time runs slower in a gravitational field, it is unable to tell us *why* time runs slower in such a situation, because it is incapable of revealing the mechanism of this *why* (this inability, being a well-known or thorny problem, has been baffling many brilliant physicists for a long time; that is, this inability is a tacitly admitted *fact* in reality. Such an inability is, of course, also a clear or undeniable *fact* in truth). More specifically to be even more noticeable and impressive, as pointed out earlier in this book, one could clearly perceive and explicitly realize, even merely from the perspective of general relativity itself, the *fact* that general relativity is really unable to solve the most fundamental problem in front of itself—*why* space and time are variable thus relative in a gravitational field. Why? How? The following simple and clear method is workable and quite helpful. If general relativity had been able to solve this most fundamental problem, its fundamentally indispensable postulate of 'invariant scales of length and time', which says that the scales of length and time at different points over an entire gravitational field are the same, would not have been necessary at all. And so, the irrefutable or undeniable naked *truth* that general relativity indispensably and desperately necessitates this postulate can enable one to perceive this *fact* clearly and realize it explicitly; this irrefutable or undeniable naked *truth* is also a clear demonstration or specific reminder that one cannot deny this *fact*. (Related clarification or reminder: all the observational or experimental tests of general relativity don't and can't change the *fact* that general relativity is really unable to solve the most fundamental problem in front of itself, *why* space and time are variable thus relative in a gravitational field. This is because there is such important general knowledge or common sense in science: observational or experimental tests themselves have neither the function nor the ability to answer the questions about the *whys* or solve the problems of the *whys*, so the task or purpose of observational or experimental tests is not to deal with these questions or these problems at all; instead, the task of answering these questions or solving these problems is, or is supposed to be, responsible by theories.) Moreover, this inability of general relativity has become further clear, through a sharp contrast, *after* knowing about how to solve *this* most fundamental problem with a new theory, which has been concisely introduced in the later part of chapter three.

In fact, this inability of general relativity turns out to be also the deepest reason or root cause that results in the serious inconsistency of general relativity and quantum mechanics, as analyzed and dug out in chapter one (for the detailed or more information about this serious inconsistency and this deepest reason, please go back to chapter one if necessary). In other words, this serious inconsistency, being a fully recognized, well-known *fact* in physics for more than half a century—such a fact, needless to say, is also an irrefutable *fact*, turns out to be a concrete confirmation or constant reminder of this inability, which can substantially help one face this inability calmly and rationally in effect. All in all, the above-mentioned pieces of specific and explicit evidence collectively and clearly point to: this inability of general relativity turns out to be a clear *fact*, an irrefutable *fact*, also an undeniable *fact*. This inability has no choice but to show us the plain fact that general relativity cannot be correct. (Related question and answer: how could one grasp this plain fact explicitly and effectively?

WHY DIFFICULT TO UNDERSTAND "THE ORIGIN AND FATE OF THE UNIVERSE"

Answer: this could be easily realized via the simple and clear thinking like the following. If a theory turns out unable to solve the most fundamental problem in front of itself, clearly, even obviously, also undeniably, the theory cannot be fundamentally correct in fact; if a theory cannot be fundamentally correct, undoubtedly, the theory cannot be correct in truth. That is, it is neither rational nor correct, neither true nor valid to say a theory, which doesn't have the ability to solve the most fundamental problem in front of itself, is a correct theory.)

Another reality arises from that general relativity is crucially based on special relativity. The analysis and discussion in chapter two have uncovered or shown such a basic fact: the two core or key concepts of special relativity, length contraction and time dilation (respectively for interpreting length becomes shorter and time runs slower at high speed), turn out to be factually incompatible thus essentially contradictory—they would directly deny each other if met together. This basic fact explicitly tells us such a solid fact: length contraction and time dilation turn out to be clearly and seriously self-contradictory. This solid fact has no choice but to tell us such a clear fact: special relativity turns out to be clearly and seriously self-contradictory. This clear fact has no choice but to show such an unavoidable fact: special relativity turns out unable to be correct. This unavoidable fact has no choice but to reveal such an inevitable fact: general relativity turns out unable to be correct, because special relativity is the crucial theoretical foundation of general relativity. Therefore, from the angle of special relativity, one can also clearly see the plain fact that general relativity cannot be correct. (Commentator: it is rational and fair to believe that all knowledgeable physicists clearly know the *fact* that general relativity is crucially based on special relativity; that is, no real or qualified experts in physics want to deny this *fact*. Moreover, it is rational and safe to conclude that all real or qualified scientists are well aware of such a self-evident or irrefutable *truth* in science: any theory, as long as it is crucially dependent on another theory that turns out to be clearly and seriously self-contradictory thus unable to be correct, cannot be correct accordingly! This self-evident or irrefutable *truth* is, of course, completely applicable to the theory of general relativity.)

With and through the two inescapable realities reviewed above, one can clearly see the revealed plain fact that general relativity cannot be correct. (Related question and answer: quite obviously, either of the two realities presented above is sufficient to show the plain fact that general relativity cannot be correct, why have you mentioned both of them? Answer: in order to strengthen the confidence of dear readers in this plain fact, I have taken the action of going the extra mile.) (If some people, especially some of those respected professionals in physics, have not been psychologically ready to face this plain fact yet, please think over the explicit *fact* that Hawking has clearly mentioned in his *A Brief History of Time* (1998 version, P. 12): general relativity and quantum mechanics cannot both be correct because they are known to be inconsistent with each other. Quite reasonably, such an explicit fact can substantially remove or reduce one's psychological hesitation towards this plain fact. Moreover, it's both rational and safe to say that all the knowledgeable professional people in physics know this explicit fact quite well, because it directly comes from and is clearly determined by the serious inconsistency of general relativity and quantum mechanics; because this serious inconsistency is so well known that it has already become very important general knowledge or common sense in physics. Thus, it seems both reasonable and fair to have the reasoning that this explicit fact can enable one to be well psychologically prepared to face this plain fact. After having this preparation, it seems not difficult that one can easily grasp this plain fact, because it has been shown by hard evidence. In addition, it seems to be a rational and realistic estimate that some, even many, sharp readers are well able to have the direct thinking like: since this explicit fact comes from the serious inconsistency of general relativity and quantum mechanics; and since the deepest reason or root cause of this serious inconsistency turns out to be due to the *reality* that general relativity is unable to solve the most fundamental problem in front of itself—*why* space and time are variable thus relative in a gravitational field, the existence of this plain fact thus becomes pretty reasonable; that is, the revealing of this plain fact is nothing to be surprised about at all! Quite noticeably, such a direct thinking can enable one to face and grasp this plain fact rationally, quickly and easily.)

Anyway, to cut a long story short, the number one obstacle (which hinders one from really understanding the idea of the expanding universe) turns out to be the revealed plain fact that general relativity cannot be correct. This plain fact has no alternative but to tell us such a basic fact: this idea cannot be correct, simply because its theoretical basis is general relativity. (The basic or noticeable nature of this basic fact is: this idea could not be really understood—either in essence or in truth or in both, because in science, it is self-evident or clear enough that rational people with reasonable thinking could not really understand an idea that turns out unable to be correct. That is to say, it turns out to be neither rational nor realistic if one claims that he or she could really understand the idea of the expanding universe. Of course, no one can rule out the possibility that some, even many, insightful readers might have such a viewpoint: even some people claim that they could really understand an idea that turns out unable to be correct, this sort of so-called "understanding" is actually meaningless or pointless in science. This viewpoint seems to be reasonably acceptable to most people, even if not everyone.) This basic fact, in turn, shows us such a clear fact: it turns out to be quite reasonable and fair that one could not really understand the hot big bang model, because the idea of the expanding universe is the main theme of this model—the major work of this model is a specific description of this idea, as pointed out earlier. This clear fact, in turn, manifests such a solid fact: it turns out to be rather rational and fair, or at least not irrational at all, that one has had great difficulty understanding the origin and fate of the universe described in *A Brief History of Time*, because this origin and fate is attached onto this model, as mentioned above. (Narrator: so, eventually the number one obstacle that prevents one from truly understanding the origin and fate of the universe in *A Brief History of Time* is ultimately ascribed to the revealed plain fact that general relativity cannot be correct.)

Up to here, if you have clearly seen and perceived the number one obstacle of *why* it's difficult to comprehend the origin and fate of the universe said in *A Brief History of Time*, please accept my congratulations. If you have not fully perceived and realized this number one obstacle yet, no worry, no hurry: the second obstacle, which prevents one from really understanding the idea of the expanding universe thus the origin and fate of the universe described in *A Brief History of Time*, is prepared to give you a hand.

The second obstacle is that the measured (cosmic) microwave radiation turns out unable to be the valid evidence for the idea of the expanding universe (please see chapter three for the detailed or more information about this obstacle). As analyzed and pointed out in chapter three, the *prerequisite* for the measured microwave radiation to be the valid evidence for this idea is that the proposed big bang was the *only* source of the microwave radiation measured today. In other words, the real implication of this *prerequisite* is that in the numerous, virtually or almost incalculable, galaxies of the vast universe, no celestial bodies can be the sources (or can be the significant sources) of the microwave radiation at present. (This is a bit like that, an environmental agency is monitoring a specific chemical compound that is very harmful to our health, and a chemical factory, factory HHH—let us say, could give off this compound more than one hundred years ago. Today, the agency monitors a high concentration of this compound in the waters of a large lake. In order to get the valid conclusion that this compound was from factory HHH, the agency has to ensure that no other factories can emit this compound at present, or at least can make sure that no other factories can be the significant sources of this compound nowadays.)

This *prerequisite*, however, cannot be valid at all, because the available information on gamma ray bursts and ultrahigh-energy cosmic rays utterly overthrows (invalidates) this prerequisite. The observed features of gamma ray bursts have revealed such a fact: in the vast universe, there are *numerous* celestial bodies that can be the sources of gamma ray bursts at present; and these celestial bodies can also be the important or significant sources of the microwave radiation measured today. And the monitored features of ultrahigh-energy cosmic rays have also shown a similar fact (that is, there are *numerous* celestial bodies in the vast universe that can be the sources of ultrahigh-energy cosmic rays at the present time; and these celestial bodies can also be the important or significant sources of the microwave radiation measured currently). Therefore, these known features, believe it or not, have evidently overthrown this *prerequisite* thus make it invalid. This invalidation, in turn, demonstrates that the measured microwave radiation cannot be the valid evidence for the idea of the expanding universe.

Undoubtedly, this demonstration substantially reinforces the conclusion obtained above from the angle of the number one obstacle, in view of the fact that the measured microwave radiation has been regarded as an important piece of evidence for the idea of the expanding universe. This reinforcement, needless to say, can further strengthen one's confidence in this conclusion. (Commentator: with these two obstacles as impassable barriers, it turns out to be both rational and understandable that one could not really understand the idea of the expanding universe, thereby the hot big bang model, thus the origin and fate of the universe interpreted and described in *A Brief History of Time*.)

At this moment, and having seen the two impassable obstacles discussed above, if you have understood *why* it's difficult to understand the origin and fate of the universe presented in *A Brief History of Time*, please accept my congratulations again. If you have not totally understood this *why* yet, still no worry, no hurry: the third obstacle, which prevents one from really understanding the idea of the expanding universe thus the origin and fate of the universe said in *A Brief History of Time*, is coming to help you. And more encouragingly, the help from the third obstacle is much greater than that from the second obstacle.

The third obstacle, as uncovered and pointed out in chapter three, is that the Doppler effect turns out unable to provide the valid observational evidence for the idea of the expanding universe, because the *prerequisite* for the Doppler effect to be able to provide the valid observational evidence for this idea cannot hold at all. (Because the detailed information about this obstacle has been presented in chapter three, here only the key points are concisely mentioned as a quick review.)

As analyzed and pointed out in chapter three, the ideal prerequisite for the Doppler effect to be able to provide the valid observational evidence for the idea of the expanding universe is to ensure that the entire redshift of the light radiated from the observed stars is due to they are moving away from us. However, because the Doppler effect *on its own* is unable to tell apart whether a redshift of light is due to a star moving away from us or due to a gravitational field, this prerequisite thus becomes that the redshift due to gravitational fields (that is, gravitational redshift) is so small that it can be negligible, compared to the redshift due to the star moving away from us. In other words, this *prerequisite* is, or simply becomes, that gravitational redshift must be insignificant! (So if this prerequisite holds, one can come to the conclusion that the Doppler effect is able to provide the valid evidence for this idea; on the contrary—if this prerequisite cannot hold, such a conclusion cannot be drawn of course.) As a result, whether or not the Doppler effect can provide the valid observational evidence for this idea hinges on whether this prerequisite holds or not (that is, whether it is valid or not, true or false). This prerequisite, nevertheless, turns out unable to hold in truth, because the available information from the observational results over the last several decades has sufficiently and clearly shown that this prerequisite cannot be valid at all.

Specifically, this *prerequisite* can be directly disproved by each or all of the four factors, actually the four known facts that have been presented in detail in chapter three (thus for the detailed information about these factors or facts, please see chapter three; it seems that there is no need to have too much detailed or repeated information here). Factor one: since the sun, even the much smaller earth, can cause significant gravitational redshift (being the observational results in the early 1960s), why cannot other even more massive stars? So this factor is sufficient to show that this *prerequisite* cannot be valid at all. Factor two: the location of the sun and the earth in the Milky Way galaxy determines that, at the locations where reside many observed stars (especially those "interpreted" as the observational evidence for the expanding universe) in many other galaxies, the gravitational fields can be much stronger than at the location of the sun and the earth; thus the resulting gravitational redshift can definitely be significant. So this factor clearly indicates that this *prerequisite* cannot be valid either. Factor three: the gravitational redshift due to dark matter definitely cannot be insignificant, because there is a large amount of dark matter in the universe—it is estimated that the amount of dark matter is about 5.5 times of that of ordinary matter (*ordinary matter is the matter like stars and planets). Therefore, the available information about dark matter shows and determines that this *prerequisite* cannot be valid again. Factor four: the gravitational redshift due to black holes cannot be insignificant, because there are numerous and massive black holes in the vast universe. As a result, the available information from black holes clearly tells us that this *prerequisite* cannot be valid once again. All in all, these four factors together have explicitly demonstrated, with more than sufficient hard evidence, such a clear conclusion, actually such a definite fact: the *prerequisite* that gravitational redshift must be insignificant cannot hold at all (that is, this prerequisite turns out to be false).

This definite fact has no choice but to tell us such an inescapable reality: it turns out that the Doppler effect cannot provide the valid observational evidence for the idea of the expanding universe (accordingly, this idea cannot be valid—either in essence or in truth or in both). This inescapable reality crucially and decisively consolidates the conclusions drawn above from the angles of the first two obstacles, corresponding to the crucial and leading role of the Doppler effect in this idea. (As mentioned in chapter three, the idea of the expanding universe originally came from the application of the Doppler effect in the 1920s, due to the American astronomer Edwin Hubble. The result obtained from this application is thus referred to as Hubble's law in today's professional materials. Therefore, the Doppler effect, being the pioneer of this idea, did play a crucially important role for the birth of this idea.) With such a crucial and decisive consolidation, it seems not difficult that one can further realize: indeed, it turns out to be rather rational thus quite understandable that one could not really understand the idea of the expanding universe.

While reading up to here, and at this moment, one can clearly see that there are three undefeatable great obstacles to one's understanding the idea of the expanding universe. They are: the theoretical basis of this idea, general relativity, turns out unable to be correct, thus this idea cannot be correct in fact; the measured microwave radiation turns out unable to be the valid evidence for this idea, which means that the validity of this idea is highly questionable in reality; the Doppler effect turns out unable to provide the valid observational evidence for this idea, consequently, also unavoidably, this idea cannot be valid in truth. (Commentator: moreover, when these three obstacles are considered comprehensively and simultaneously, one can easily perceive or realize that they are clearly consistent with each other. Quite obviously, also rather rationally, such a clear consistency not only considerably strengthens the validity and reliability of the separate role of each individual obstacle, but also substantially intensifies and increases the overall effect of all these obstacles. These mutually consolidating effects can enable one to see more clearly the conclusions we have obtained above.)

Altogether, these three insurmountable great obstacles have clearly shown us a fact that cannot be clearer: it is actually impossible—in fact, it is neither rational nor realistic to imagine that one could really understand the idea of the expanding universe. This crystal clear fact, in turn, displays such a manifest fact: it is not possible that one could really understand the

WHY DIFFICULT TO UNDERSTAND "THE ORIGIN AND FATE OF THE UNIVERSE"

hot big bang model (because its main theme is this idea). This manifest fact, in turn, inescapably reveals such a candid *reality:* it turns out to be both reasonable and realistic that one could not truly understand the origin and fate of the universe described in *A Brief History of Time* (because this origin and fate is determined by the hot big bang model—please remember that this origin and fate is attached onto this model). Or stated frankly, the real implication of this candid reality is actually or simply becomes: it would be really and truly weird if one claimed that he or she could truly understand this origin and fate. As a result—as an explicit and pleasant result in effect, this candid reality is a great relief or consolation for the people who might have felt discouraged or insipidly disappointed for having not truly comprehended this origin and fate. (Related question and answer: can this candid reality be changed by the other models or proposals mentioned in *A Brief History of Time?* Answer: the answer to this question is the very topic to be dealt with immediately, as the effort of going the extra mile on the issue of this candid reality.)

What possible reactions may arise from the candid reality revealed above (which is that it turns out to be reasonable and fair that one could not really understand the origin and fate of the universe described in *A Brief History of Time*)? Some dear readers may raise the question like: can such a reality be changed by the other models or proposals mentioned in *A Brief History of Time*, since Hawking spent about 20 pages talking about them in it (P. 126 ~ 146, 1998 version)? Others might psychologically feel that this reality, though candid, is somewhat too harsh in their minds, thus would hope that these other models or proposals could buffer or mitigate such a "harsh" reality.

These possible reactions seem to assign us an extra task: it's better to inspect or analyze whether these other models or proposals (mentioned in *A Brief History of Time*) can change the reality that it turns out to be both rational and realistic that one could not really comprehend the origin and fate of the universe presented in *A Brief History of Time*. In addition, the conclusion from such an inspection may help more people see this reality more clearly. Thus, as the action to complete this extra assignment, also as an action of going the extra mile on the main subject of this chapter, we are about to do this inspection.

Fortunately, at this moment, it is no longer difficult to carry out this inspection, because we have had a clear and definite principle to guide our inspection. This principle, as the abbreviation or epitome of the results discussed above, is: the major premise of the origin and fate of the universe interpreted and described in *A Brief History of Time*, the hot big bang model, has utterly collapsed; that is, this major premise turns out to be false, simply because what this model describes is an idea (the idea of the expanding universe) that turns out unable to be correct. (Commentator: if some people, especially some of those respected professionals in physics, are still psychologically hesitant or reluctant towards the revealed or shown fact that the idea of the expanding universe turns out unable to be correct, please closely notice the three clear *facts* that forever linger around and hover above this idea. First fact: the theoretical basis of this idea is the theory of general relativity; however, it turns out that this theory cannot be correct, as pointed out earlier in this chapter. Second fact: this idea requires the indispensable support from the measured microwave radiation; nevertheless, it turns out that the measured microwave radiation cannot be the valid evidence for this idea, as reviewed above. Third fact: this idea needs the crucial and decisive support from the Doppler effect, but it turns out that the Doppler effect cannot provide the valid observational evidence for this idea, as just mentioned above. Quite obviously, these three clear *facts* can substantially help one remove or reduce his or her psychological hesitation or reluctance towards this revealed or shown fact, which is a considerable help for him or her to face and realize this revealed fact calmly, rationally and effectively.) More specifically thus more explicitly, this revealed fact (which is that the idea of the expanding universe turns out unable to be correct) has no choice but to show us another clear and inevitable fact: the hot big bang model cannot be correct, simply because this idea is exactly the main theme of this model, as clearly mentioned several times above. That is, this guiding principle literally becomes that the hot big bang model cannot be correct.

Such a guiding principle, in turn, also in fact, determines and demonstrates the following two clear and objective criteria or standards that are easy to work with in this inspection. One is that, no matter whether these other models or proposals (mentioned in the book *A Brief History of Time*) make sense or not, because their major premise is still the hot big bang model, cannot significantly influence the origin and fate of the universe interpreted and described in that book in truth. Another is that, whether or not these other models or proposals can be smoothly connected to the critical rate of expansion in the hot big bang model, because their entire paradigm or stereotype is still this model, actually cannot help us understand this origin and fate in essence.

Having had this guiding principle and the two clear standards from it, now we get prepared to work on our specific task: to inspect whether these other models or proposals, which were mentioned in the book *A Brief History of Time*, can change the reality that it turns out to be both rational and realistic that one could not really comprehend the origin and fate of the universe described in that book.

Let us first inspect the "no boundary" proposal suggested by Stephen Hawking. Briefly, this proposal is about the possibility or conception that the universe was finite but had no boundary or singularity in the very early stages; that is, the initial state of the universe was finite in extent but had no singularity that outlined a boundary or edge. On the whole, since the paradigm or major premise of this proposal is still the hot big bang model, thus this proposal doesn't and can't help us understand the origin and fate of the universe described in *A Brief History of Time* at all. In other words, because the general direction of this proposal turns out unable be correct, this proposal doesn't and can't have the ability to help us understand this origin and fate in truth—stated plainly to be exact.

Specifically, please closely notice that the theoretical foundation of this proposal is the assumed quantum theory of gravity (this theory is also called quantum gravity theory or simply as quantum gravity). Since this assumed theory has not existed yet, we need to check on the *prerequisite* for its existence, in order to be aware of the possibility to have it in the future. (If this *prerequisite* exists, we can have the conclusion that this assumed theory may exist someday. However, if it turns out that this *prerequisite* cannot exist at all, then this assumed theory will definitely not exist, either in the near future or in the distant future.) Then what is this crucial *prerequisite?* And can it exist? Let us find the answers to these questions.

Given that the most important purpose of the assumed quantum theory of gravity is to combine general relativity and quantum mechanics, clearly the *prerequisite* for this assumed theory to exist is that general relativity and quantum mechanics have to be consistent or compatible with each other (an inevitable implication of this prerequisite is: if these two theories are known to be inconsistent with each other, there must be a way to solve this inconsistency). However, as analyzed and pointed out in chapter one, general relativity and quantum mechanics are seriously inconsistent with each other (about this serious inconsistency, please go back to chapter one if necessary). Moreover, this serious inconsistency has not been, cannot be, and will never be solved within the paradigm or stereotype of either general relativity or quantum mechanics or both, being determined by the deepest reason of this serious inconsistency.

This deepest reason, having been dug out in chapter one, turns out to be: general relativity is unable to solve the most fundamental problem in front of itself, *why* space and time are variable thus relative in a gravitational field, simply because it doesn't have the ability to reveal the mechanism behind this *why;* and quantum mechanics cannot solve the most fundamental problem in front of itself, *why* there are quantum states, simply because it is utterly incapable of revealing the *quantum* mechanism of why and how photons, being the tiny and discrete quanta or particles of light, get their velocity c (the speed of light) from the electron that emits them. (Related questions and answer: could one clearly perceive and explicitly realize, even merely from the large perspective or basic features of quantum mechanics itself, the *fact* that quantum mechanics is really unable to solve the most fundamental problem in front of itself—*why* there are quantum states? How? Answer: yes, and certainly! Either of the following two simple and clear methods is workable and highly effective. First method: if quantum mechanics had been able to solve this most fundamental problem, its core *assumption*, or the fundamentally indispensable *assumption* for the birth of quantum mechanics, which says that all forms of energy released from electrons are in the way of discrete units or bundles called quanta (being often referred to as photons nowadays), would not have been necessary at all; thus this core *assumption* would never have appeared. And so, the irrefutable and unavoidable *reality* that quantum mechanics indispensably and desperately necessitates this core *assumption* can enable one to perceive this *fact* clearly and realize it explicitly; this irrefutable and unavoidable *reality* also shows and determines that no one is able to deny this *fact*. Second method: quantum mechanics is *wave* mechanics—none of the related professional people in physics deny the *truth* that quantum mechanics is wave mechanics, because they know this *truth* quite well. However, *wave* mechanics doesn't and can't have the function or ability to solve the problem of why there are *quantum* states at all. Therefore, this *truth* can enable one to perceive this *fact* more clearly and realize it more explicitly.) (Related reminder or clarification: as clearly mentioned in chapter one, being important general knowledge or common sense in science, the task or purpose of experimental tests is not to answer the questions about the *whys* or solve the problems of the *whys*, because experimental tests themselves have neither the function nor the ability to do that; the task of answering these questions or solving these problems is, or is supposed to be, responsible by theories instead. Thus this general knowledge or common sense clearly and explicitly tells us that all the experimental tests of quantum mechanics don't and can't change the *fact* that quantum mechanics is really unable to solve the most fundamental problem in front of itself—*why* there are quantum states. In other words, this general knowledge or common sense can certainly make or help one see this *fact* even more clearly and explicitly, thus grasp it even more tangibly and definitely.) So it is the fundamental nature of this deepest reason that demonstrates and determines that general relativity and quantum mechanics are essentially inconsistent or inherently incompatible with each other, which unavoidably points to or explicitly shows that there is utterly no way to solve the serious inconsistency between these two theories. Consequently, there is neither correct nor successful way to combine them in fact; that is, these two theories cannot be put in the same package at all. Therefore, we have to draw the solid conclusion that the *prerequisite* for the existence of this assumed theory does not exist at all.

What does this solid conclusion tell us? It tells us such a hard fact: the assumed quantum theory of gravity cannot exist at all (that is, it not only hasn't existed, but also doesn't and won't exist). This hard fact is plainly equivalent to such an inescapable truth: Hawking's "no boundary" proposal turned out to be built upon a theory that is definitely impossible to exist! So by analogy, this proposal turns out to be the moon in the water, the flowers in a mirror. Such a proposal, of course, cannot change the *reality* that it turns out to be reasonable thus understandable that one could not really understand the origin and fate of the universe described in *A Brief History of Time*. (Related question and answer: towards the hard fact that the assumed, actually the claimed, quantum theory of gravity is definitely impossible to exist, some people, especially some of those respected professional people in physics who are now working on this theory, may feel a surprise at first sight; what is an effective way to remove or alleviate such a surprise then? Answer: the conception of the so-called *quantum* theory of gravity was initially "inspired" from quantum mechanics. However, it should be carefully noticed that quantum mechanics itself is even unable to solve the very problem of *why there are quantum states*, the most fundamental problem it has to face, as just mentioned above. In fact, this inability of quantum mechanics has been perplexing many brilliant physicists for a long time. That is, this inability, being a tacitly admitted *fact*, is either an explicitly realized *fact* or an actually recognized *fact*; such a fact, needless to say, is also an irrefutable or undeniable *fact*. So, this inability is simply equivalent to the plain truth that this sort of inspiration cannot be truly valid in essence. Quite obviously, such a plain truth can effectively remove or alleviate one's surprise towards this hard fact.) (If you have not fully grasped this hard fact yet, no worry, no hurry: the coming information will give you a hand.)

Moreover, the hard fact revealed above (which is that the assumed quantum theory of gravity is definitely impossible to exist), could be seen more clearly when it is pasted onto the background of the "harsh" actuality this assumed theory has to face. How harsh is this actuality? Let us see facts! Let facts tell us!

As admitted in some highly authoritative resources (for example, the encyclopedia of physics; and *Science* Magazine, the issue of July 8, 2005, to be exact—again, this magazine is widely known to be one of the most authoritative, most influential and most famous publications in science), the assumed or presumed quantum theory of gravity is one of the most difficult intellectual challenges theoretical physicists have ever faced, also one of the most difficult puzzles in science. Consequently, this presumed theory, albeit an ongoing program *for more than half a century* as the hottest research area in theoretical physics, has not yet borne fruit—has not even seen a faint hint of real success in fact; needless to say, this presumed theory hasn't existed so far. (Commentator: thus both in fact and in reality, it is quite safe to say that all the related professional people in physics, particularly those physicists who have been working on this unsuccessful theory over the last several decades, are very familiar with this kind of long-standing and formidable challenge or difficulty.)

Where does such an enormous challenge or difficulty come from? And what is the deepest reason incubating the formidable challenge or difficulty of this kind? It has been clearly identified and fully recognized that this kind of challenge or difficulty comes from the well-known *fact*, which is that general relativity and quantum mechanics are seriously inconsistent or crucially incompatible with each other. That is, the serious inconsistency or crucial incompatibility of these two theories has been explicitly known to be the most fundamental reason or root cause leading to this kind of long-standing and formidable challenge or difficulty. Moreover, this serious inconsistency or crucial incompatibility has even been openly admitted as the deepest disaster of modern physics. Regrettably, in the face of this deepest disaster, it seems that not many related scientists, for whatever the reasons, have touched the deepest reason behind this deepest disaster. Then what is this deepest reason?

This deepest reason, as just mentioned above (also as analyzed and pointed out in chapter one), turns out to be the two fundamentally important, also inescapable, facts. One is that general relativity doesn't have the ability to solve the most fundamental problem in front of itself—*why* space and time are variable thus relative in a gravitational field; another is that quantum mechanics cannot solve the most fundamental problem in front of itself—*why* there are quantum states. In other words, this deepest reason is that neither general relativity nor quantum mechanics is able to solve the most fundamental problem in front of itself. Consequently, also inescapably, this deepest reason turns out to be the impassable barrier to constructing the presumed quantum theory of gravity. Therefore, believe it or not (admit it or not), the fundamental or inherent nature of this deepest reason sufficiently demonstrates and determines that this presumed theory has not been, cannot be, and will never be born within the paradigm or stereotype of either general relativity or quantum mechanics or both. (By contrast, after discovering the new theory that has solved the very problem of *why* space and time are variable thus relative in a gravitational field—this new theory has been briefly mentioned in the later part of chapter three; and after finding out the other new theory that has solved the very problem of *why* there are quantum states—this new theory will be concisely introduced in chapter six as a newly discovered physical law, these two new theories are totally consistent and compatible with each other. So, through this sharp contrast, one can see and/or perceive this deepest reason even more distinctly and definitely.)

All in all, with and through these facts and contrasts, one can not only see more clearly the hard fact already revealed above (which is that it is definitely impossible for the assumed quantum theory of gravity to exist), but also distinctly realize that this assumed theory hasn't been, can't be and won't be born within the paradigm of either general relativity or quantum mechanics or both. This realization further concentrates the gist extracted above: really and surely, Hawking's "no boundary" proposal turned out to be built upon a non-existent theory; indeed this proposal turns out to be the moon in the water, the flowers in a mirror, or castles in the air. (Related question and answer: are there some available clues that can explicitly remind us of the uncovered fact that this proposal turns out to have been built on a non-existent theory, the assumed quantum theory of gravity? Answer: yes, there are. For example, this kind of clues includes that this proposal had already incubated the model that was forced to fabricate *imaginary* time with *imaginary* numbers. Who can really and truly understand the so-called *imaginary* time?!? Perhaps nobody can—to be rational and realistic, to be objective and fair. As a result, the essentially incomprehensible *imaginary* time turns out to be actually a clear and noticeable clue pointing to that this proposal has been built on a non-existent theory, the assumed quantum theory of gravity; particularly after it has been shown, with hard evidence, that this assumed theory doesn't exist at all. So, with this clear clue, it seems quite reasonable and fair that some, even many, insightful and sharp readers can rationally perceive: the so-called *imaginary* time, when viewed from the angle of comprehension, could be merely some kind of "compensation" for the actually *non-existent* quantum theory of gravity. Such a perception is obviously an effective help for these readers to see this hard fact more clearly and explicitly, thus grasp it more realistically and definitely.)

So, personally I really couldn't figure out why Hawking suggested such a proposal based on the assumed quantum theory of gravity, because it seems quite safe to believe that he must or should have known that this assumed theory requires the combination of general relativity and quantum mechanics; and because he should have realized that this combination cannot give rise to a correct theory. Why? Please notice that Hawking has also clearly mentioned in *A Brief History of Time* (1998 version, P. 12): general relativity and quantum mechanics cannot both be correct, because these two theories are known to be inconsistent with each other. (Commentator:

there is a clear, also self-evident, rule in science, which is that *any* theory, as long as it requires the combination of two theories that cannot both be correct, cannot be correct at all.) Personally, I also greatly admire the frankness of Hawking about his own work. He has admitted that it is impossible to test his (this) proposal and/or the model based on it, at least in the near future.

After finishing the task of inspecting the "no boundary" proposal, let us turn our eyes to the other models mentioned in *A Brief History of Time*. These models included the old inflationary model (suggested by Alan Guth), the new inflationary model (jointly put forward by Andrei Linde, Paul Steinhardt and Andreas Albrecht), and the chaotic inflationary model (proposed by Andrei Linde). Let us start our work from the most important common feature of these three models.

While they were different in many details, these models had the most important common feature. This feature was: all these models targeted at connecting to the hot big bang model with possible means—by trial and error, and through various attempts to describe the very early universe (within about only one second just after the proposed big bang), including the initial state of the universe, its boundary conditions, and its initial configuration. Moreover, this common feature, being the crucial and central topic of all these models, was also their most fundamental and most important goal.

What does this most important common feature tell us? It shows and determines three decisively important things that are clear, even obvious, also unavoidable or irrefutable. First, whether these three models can be successfully connected with the hot big bang model or not, they cannot help us understand the origin and fate of the universe described in *A Brief History of Time* at all. This is because the hot big bang model itself, being the consequence of describing the idea of the expanding universe that turns out unable to be correct, turns out unable to be correct in truth, as revealed and pointed out above. Second, while professor Hawking analyzed and mentioned many "advantages" or "strong points" of these models in his *A Brief History of Time*, all these "merits" turn out unable to be valid in fact—at least cannot be essentially valid for sounding less harsh, because they are all with respect to the main theme of the hot big bang model, the idea of the expanding universe. Third, even though Hawking also pointed out some flaws of these models, such as the problem that the bubbles in the old inflationary model could not join up, and the dilemma that the bubbles in the new inflationary model were bigger than the size of the universe, all these flaws turn out unable to be valid in truth, because they are all from the standpoint of the hot big bang model—the criterion point of all these flaws is still based on what this model describes, the idea of the expanding universe. All in all, either individually or collectively, none of these models can change such a *reality:* it turns out to be quite reasonable thus pretty understandable that one could not really understand the origin and fate of the universe interpreted and described in *A Brief History of Time*.

In addition, it should be pointed out or noticed that the so-called anthropic principle (even its weak version), though mentioned many times in *A Brief History of Time*, still doesn't and can't give any help in "explaining" the origin and fate of the universe said in *A Brief History of Time*. This is because such a sort of "explanations", after taking off its disguise or camouflage, is actually like the following simple situation. To the question: why did professor Hawking write the book *A Brief History of Time?* Answer: because it has existed; if he had not written it, then it would not have existed, thus no one would ask this question. (Commentator: consequently, the so-called anthropic principle doesn't and can't affect the *reality* that it turns out to be quite reasonable thus totally comprehensible that one could not really comprehend the origin and fate of the universe described in *A Brief History of Time* either.)

Last but not least, before ending this chapter, I have a fundamentally important question to discuss with dear readers, because the discussion on this question can substantially help one perceive and realize the key point of this chapter more clearly and definitely (this key point is: it turns out to be rather rational thus quite understandable that one could not really understand the origin and fate of the universe described in *A Brief History of Time*). To the theoretical basis of the hot big bang model, *general relativity*, the inescapable and crucial question is: *do the scales of space and time in the universe change?* If the answer is YES, then general relativity admits itself is wrong, because it has the postulate of 'invariant scales of length and time', which claims that the scales of length and time at different points over an entire gravitational field are the same (note: the universe, apart from a tiny fraction of space occupied by various celestial or heavenly bodies themselves, is composed of the numerous gravitational fields around the numerous celestial bodies such as stars, planets and black holes in each galaxy, as well as the almost or virtually incalculable gravitational fields around the almost or virtually incalculable galaxies). If the answer is NO, this answer turns out to be the direct cause of general relativity being unable to solve the most fundamental problem in front of itself, *why* space and time are variable thus relative in a gravitational field. If the answer is I DON'T KNOW, this answer can be, at least in a sense, a clear signal that may point to that general relativity cannot be correct, because this inescapable and crucial question is fundamentally important, when viewed from the angle of the most fundamental, also the most important, subject of general relativity—this subject deals with the issues of space and time in gravitational fields and in the universe. As a result, each or any of these three answers can greatly expand one's field of vision on the key point of this chapter, which is an explicit and considerable help for one to see this key point more clearly, thus grasp it more effectively (because this key point is attached onto the hot big bang model whose theoretical basis is general relativity).

Now, one can see that this chapter has revealed the secrets of *why* it's difficult to understand the origin and fate of the universe described and interpreted in the book *A Brief History of Time*. This revealing demonstrates that the trouble or responsibility, in which people could not really understand this origin and fate, turns out not to be on the side of the people who have read that book. This demonstration has thus driven away the clouds over the heads of many people, who might have felt discouraged or disappointed for having not really understood this origin and fate, as long as they have known these secrets. This pleasant effect is the take home message of this chapter!

CHAPTER 6

THE CERTAINTY MECHANISM BEHIND THE UNCERTAINTY PRINCIPLE

[The Window on This Chapter]

In the face of the *reality* that quantum mechanics and general relativity cannot both be correct because they are known to be inconsistent with each other (*A Brief History of Time*, 1998 version, P. 12), one can't help asking: is quantum mechanics correct?

With this question kept in mind, one can't help thinking of or asking: is the present interpretation of the uncertainty principle correct (since this principle is the fundamental nature of quantum mechanics)?

Following this question, one can't help thinking over: is there and what is the *certainty* mechanism behind the *uncertainty* principle (because the present interpretation of this principle says that there is no certainty mechanism behind it)?

Then what will readers experience and enjoy in this chapter?

After reading this chapter, dear readers will clearly see, though perhaps somewhat out of the blue: aha! It turns out that there is the *certainty* mechanism behind the *uncertainty* principle (no wonder it's hard to understand this principle according to the interpretation that there's no certainty mechanism behind it).

The wonderfully fascinating *certainty* mechanism behind the famous *uncertainty* principle is coming towards us—with strong and steady strides!

This spectacular *certainty* mechanism is about to be unfolding itself right in front of us—with splendid and irresistible charm! Dear readers, are you ready?

Nearly ninety years ago, in 1927—to be exact, suddenly, and somewhat mysteriously, appeared a marvelous, large building in the world. The appearance of this spectacular structure caused a worldwide sensation very soon. Since then, this splendid building has been getting more and more famous; as time went on, and by today, it has become a celebrated edifice. At first sight, what catches the eyes of people is its impressive size. From the top of this majestic building, one can look down at, through a bird's-eye view, all other structures nearby—with binoculars, people can even enjoy the beautiful scenery of the entire city and its outskirts. And this building is so tall that a warning signal has to be installed to avoid planes striking it. However, what really makes this large building more famous and attractive actually lies with the big puzzle remained: does it have a foundation? Or is there a solid foundation beneath it?

This tantalizing puzzle has been perplexing many curious people since the birth of this large building. Around and over this big puzzle, has been lingering and hovering an intense dispute from two vigorously disagreeing camps. One camp has been persisting in: there is no foundation at all beneath this building, because no one has ever found out the foundation. Another camp has been insisting on: there must be a solid foundation beneath such a tall and large building, because if not, it would have already collapsed. Moreover, what makes this puzzle further tantalized is that the argument from either camp sounds equally reasonable, particularly from its own standpoint.

With the title of this chapter as an explicit reminder, and by what has been clearly seen through [The Window on This Chapter], along with the noticeable help from the information in the above two paragraphs, it seems not difficult that one can perceive or realize that this large building represents the famous uncertainty principle, and whether there is a solid foundation beneath this building stands for whether there is the *certainty* mechanism behind the *uncertainty* principle. Yes! Such a perception or realization is correct!

With such a perception or realization, one can easily think of or ask the related questions like: is it of really extreme necessity or truly remarkable importance to inspect whether or not there is the *certainty* mechanism behind the *uncertainty* principle? Or is it really fundamentally important to dig out this *certainty* mechanism? Does it surely have profound implications and far-reaching significance to reveal this *certainty* mechanism? The answer to all these questions is: YES. Why? Let reality and fact tell us! And let us see reality and fact (through the fully recognized, also fundamentally profound and extensively influential, inconsistency of quantum mechanics and general relativity)!

In facing the *reality* that quantum mechanics and general relativity cannot both be correct because of their inconsistency with each other (*A Brief History of Time*, 1998 version, P. 12), one can naturally, or easily, think of the question: is quantum mechanics correct? With this question as a clear reminder, it seems quite reasonable that one can't help asking: is the present interpretation of the uncertainty principle correct, because of the *fact* that this principle is the fundamental nature of quantum mechanics? Following this question, it seems rather rational that one can't help thinking of or asking: is there the *certainty* mechanism behind the *uncertainty* principle, because of the *fact* that the present interpretation of this principle says that there is no certainty mechanism behind it? Thus, the question of whether there is the certainty mechanism behind the uncertainty principle can be closely related to the widely acknowledged inconsistency of quantum mechanics and general relativity, a well-known inconsistency with crucially important, far-reaching and profound consequences. (Commentator: in fact, the serious inconsistency between quantum mechanics and general relativity has been noticed and recognized for more than half a century; and by today, this serious inconsistency has already become very important general knowledge or common sense in physics as a well-known *fact*, as a clear *fact*, also as an undeniable *fact* or as an inescapable *fact*. In reality, as mentioned in chapter one, this serious inconsistency has three "terrible" names at present: the deepest disaster of modern physics, the greatest challenge in theoretical physics, and one of the most fundamental problems unsolved in science. In effect, through the fundamentally important status of quantum mechanics and general relativity—both of them are the main theoretical pillars of modern physics, one can clearly and easily perceive the profound consequences of this serious inconsistency: what would happen if the main pillars of an extremely large building were seriously inconsistent with each other?!? Thus, it is no exaggeration to say that the real implications of this serious inconsistency actually have a crucial, even decisive, influence on the fate and future of modern physics in truth. And so, we have to bravely face such a serious inconsistency; accordingly, we have to earnestly deal with the question of whether there is the certainty mechanism behind the uncertainty principle.) Therefore, from the perspective of this serious inconsistency, it is of really extreme necessity or truly remarkable importance to inspect whether or not there is the *certainty* mechanism behind the *uncertainty* principle. Moreover, from the standpoint of the fundamentally profound consequences of this serious inconsistency, we cannot afford not to take on the crucially or exceptionally important task of inspecting whether there is the *certainty* mechanism behind the *uncertainty* principle.

One can think conversely if necessary: since the *reality*, again which is that quantum mechanics and general relativity cannot both be correct because of their inconsistency, appears along with the interpretation that there is no certainty mechanism behind the uncertainty principle, perhaps the discovery of the certainty mechanism behind this principle will enable one to see or realize the deepest reason or cause resulting in their inconsistency, actually their serious inconsistency, more clearly. This can also lead one to be well aware: it is really important to examine whether there is the *certainty* mechanism behind the *uncertainty* principle; and we dare not be blind to the question of whether there is the *certainty* mechanism behind the *uncertainty* principle. (Commentator: YES! Clearly knowing about this deepest reason is definitely fundamentally important, simply because the actual implications of the serious inconsistency between quantum mechanics and general relativity are crucially and closely related to the fate and future of modern physics, as just mentioned above.)

More than that, since the uncertainty principle has been prevalently regarded as one of the most significant theories in physics, also being one of the most influential concepts in science, clearly, even obviously, the *certainty* mechanism behind this principle (if this *certainty* mechanism really exists) is not only one of the most significant mechanisms in physics, but also one of the most influential and most fundamental mechanisms in science. Accordingly, also plainly and undoubtedly, it is surely of profound implications and far-reaching significance to reveal the *certainty* mechanism behind the *uncertainty* principle, if this *certainty* mechanism does exist. (Related question and answer: quite noticeably, you have gone the extra mile for emphasizing the crucial necessity and/or fundamental importance to scrutinize whether there is the *certainty* mechanism behind the *uncertainty* principle. Why? Answer: if I didn't do that, some people, especially some of those respected professional people who have been deeply ingrained into the paradigm or stereotype of saying that there is no certainty mechanism behind the uncertainty principle, might tend to blame or think that any effort spent in revealing the certainty mechanism behind this principle seems to be either looking for trouble or meaningless.) All in all, the pieces of relevant information presented above have clearly shown, or explicitly point to: it is of crucial necessity and/or fundamental importance to have a close scrutiny of whether there is the *certainty* mechanism behind the *uncertainty* principle.

(Commentator: having realized the crucial necessity and/or fundamental importance to inspect whether or not there is the *certainty* mechanism behind the *uncertainty* principle, one might curiously ask: how to show the explicit existence of this certainty mechanism? The answer can't be simpler: find it out! This is quite similar to: having found out the continent of North America is the clear demonstration that there is this continent; having found out diamond beneath a certain place is the hard evidence that there is diamond beneath this place.)

Well, the leading role of this chapter is about to be coming! The central task of this chapter is a brief introduction of how to reveal the *certainty* mechanism behind the famous *uncertainty* principle, and how to give this certainty mechanism an accurate description. The fascinating *certainty*

THE CERTAINTY MECHANISM BEHIND THE UNCERTAINTY PRINCIPLE

mechanism behind the wonderful *uncertainty* principle will be gradually and elegantly unfolding right before our eyes!

Before carrying out this central task, we need to have some basic knowledge about the uncertainty principle, like: what is its origin? What is this principle about? And what are its major features? But in order to go to the uncertainty principle, we cannot avoid quantum mechanics, since this principle is the fundamental nature of quantum mechanics; in fact, quantum mechanics is exactly the origin or root of this principle (which will be clearly seen after a short while). So let us start the journey of this chapter from briefly reviewing some background information about quantum mechanics.

Quantum mechanics came to the world in the 1920s, in 1925—to be exact. In that year, two mathematically different theories on quantum mechanics appeared independently. One was the *wave* mechanics of Erwin Schrodinger (Mathematical theory of wave mechanics); another was Werner Heisenberg's matrix mechanics (Mathematical theory of matrix mechanics). Interestingly, Schrodinger's theory caused high attention very soon because it was mathematically simple, whereas Heisenberg's theory didn't, partly because it was mathematically complicated. (This historical event has also taught us both a lesson and an experience: if one finds a new theory, particularly a radically new theory, it's better, and much better to choose a simple way to express it.) Shortly after the appearance of these two apparently different theories of quantum mechanics, Schrodinger proved in 1926 that they were totally equivalent: one could be derived from the other, and vice versa. (Friendly reminder: thus what should be clearly noticed and well aware is that the so-called quantum mechanics is actually *wave* mechanics; please don't be misled by the apparently eye-catching word 'quantum' in quantum mechanics.)

The overall picture of quantum mechanics is reflected in the following three key elements. First, its most prominent characteristic feature lies in the fact that the so-called *quantum* mechanics is *wave* mechanics. Second, its core concept lies with the *assumption* that all forms of energy are released in the way of discrete units or bundles called quanta, known as photons today. Third, its core dynamic mark rests with that only one force is involved; this force is Coulomb force (that is, electric force, being familiar to many people). As the collective function of these key elements, the task of quantum mechanics is, by using wave mechanics, to describe how electrons, under the action of electric force, radiate photons—quantum or discrete energies (which are the tiny and discrete quanta or particles of light), being the same thing as how photons are radiated or generated from electrons (of course, one photon is radiated or emitted from an electron alone). Because photons are electromagnetic radiation, this task is actually about how electromagnetic radiation is generated from electrons (which is exactly the same thing as how photons are generated or radiated from electrons). The basic feature or fundamental nature of this task determines that, in order to have an in-depth knowledge about quantum mechanics or in order to know about its history from a large perspective, the phenomenon of electromagnetic radiation is an unavoidable topic.

Historically, the phenomenon of electromagnetic radiation had played a crucially important role at the critical turning point in the development from classical physics to modern physics. In the late 19th century, the traditional wave theory had brought about such an obviously ridiculous result: the total energy radiated from a hot object or body (such as an oven or a star) through electromagnetic radiation would be infinite! To solve this problem, Max Planck put forward the quantum hypothesis about electromagnetic radiation in 1900—the year widely known as the dividing line between classical physics and modern physics, leading to the germination of modern physics. More than a decade later, in 1913 Niels Bohr came up with his famous model of the hydrogen atom, which was also about electromagnetic radiation (electrons emit photons). Although this model agreed closely with the wavelengths of the light emitted from the simplest atom, the hydrogen atom (with only one electron orbiting around its nucleus), it had obvious flaw or inability (such as unable to calculate the wavelengths of the light emitted from atoms with more than one electron). The flaw of this type led scientists to continue exploring a better method to explain the phenomenon of electromagnetic radiation in the next decade or so. This kind of exploration first led Louis de Broglie to suggest the wavelength equation for electrons in 1924. This wavelength equation then played an indispensable role for the birth of quantum mechanics in the next year—1925, because the wavelength of electrons determined with it was directly put into quantum mechanics. The theory of quantum mechanics was a noticeable improvement over Bohr's model.

Following the route above, one can clearly see two very important facts. One is that the phenomenon of electromagnetic radiation was eventually responsible for the birth of quantum mechanics (that is, this phenomenon was the cradle of quantum mechanics). Another is that the advantage of quantum mechanics was also reflected in its description of electromagnetic radiation. These two facts clearly show that electromagnetic radiation was indeed a fundamentally important phenomenon in the development of physics, given that quantum mechanics has been widely viewed as a fundamentally important theory.

However, in front of the fundamentally important phenomenon of electromagnetic radiation, the theory of quantum mechanics has to face two inescapable, also crucially important, basic facts. Let us start from the first fact. Since photons are electromagnetic radiation, and since photons are emitted from electrons, then quantum mechanics actually describes, with wave mechanics, how electromagnetic radiation is generated or emitted from the electrons that are under the action of the *only* force, electric force. As a result, this description explicitly tells us such a plain *fact:* in quantum mechanics, the so-called electro*magnetic* radiation does not include the action of *magnetic* force at all! In other words, according to the description of quantum mechanics, the action of *magnetic* force utterly evaporates, making "electro*magnetic* radiation" a purely empty terminology! (I believe that some, even many, sharp professors and students, when teaching and studying quantum mechanics, might have already thought of the question like: in its description of electro*magnetic* radiation, where does *magnetic* force go? That is to say, some, even many, brilliant or attentive professional people might have already noticed or realized the plain *fact* that quantum mechanics, in its description of electro*magnetic* radiation, really does not include the action of *magnetic* force! And so, it is actually rather rational and safe if one concludes that neither the real experts in modern physics nor knowledgeable physicists want to deny this plain *fact*, because they must/should have known it very well. All in all, this plain *fact*, being a clear and definite fact, is an irrefutable or undeniable fact in truth.) (Related question and answer: what is the profound or real implication of this plain *fact?* Answer: the explicit existence of this plain *fact* can rationally, even naturally, lead one to think of or ask the question like: is quantum mechanics really a correct theory? Clearly, even obviously, the rational existence of this kind of question can greatly reduce the psychologically painful reaction of some people, especially some of the professional people in physics, if quantum mechanics turns out to be incorrect. In this aspect, we should thank Professor Hawking for having already pointed out: quantum mechanics and general relativity cannot both be correct because of their inconsistency with each other.) (Friendly reminder: the magnetic force in the fundamentally important phenomenon of electro*magnetic* radiation will soon be presented as a newly discovered force with a new physical law in this chapter, so dear readers will see the elegant and graceful appearance of this crucially important force a short while later.)

The second basic fact also directly comes from quantum mechanics itself. Its overall picture has unmistakably displayed or clearly shown such an explicit *fact:* the so-called quantum mechanics actually uses the *continuous* wave mechanics to describe the *discrete* or *discontinuous* assumption, which says that all forms of energy are released in discrete units or bundles called quanta, or photons, as just mentioned above. (Commentator: yes, that is definitely correct! Not only that, the basic fact, which is that quantum mechanics <u>is</u> wave mechanics, is not only a clear or irrefutable fact, but has also become important general knowledge or common sense in physics, even in science. Thus, it is both realistic and safe if one has the conclusion that neither the real experts in modern physics nor knowledgeable physicists want to deny this basic fact, simply because they know it very well. And in reality, if anyone says that quantum mechanics is not wave mechanics, all other people, as long as they have some basic knowledge on quantum mechanics, will think this person does not know of quantum mechanics at all.) (Related question and answer: then what can we see from this explicit *fact?* Or what is the actual or real implication of this explicit *fact?* Answer: it seems reasonable, at least not irrational at all, that this explicit *fact* can enable some, even many, insightful and sharp readers to realize or raise the question like: how could *wave* mechanics magically become *quantum* mechanics?!? Quite obviously, the rational existence of this kind of question can be a noticeable signal pointing to: quantum mechanics may be a theory with serious flaws.)

After having obtained the background information on quantum mechanics, we are ready to go to some basic knowledge about the uncertainty principle. This, as mentioned above, includes: what is its origin? What is this principle about? And what are its major features? The uncertainty principle originated from quantum mechanics, because it was the theoretical basis on which this principle was established (this principle was developed by Werner Heisenberg in 1927 soon after the birth of quantum mechanics in 1925). This fact was clearly reflected in the specific process of how he discovered this principle.

Heisenberg discovered the uncertainty principle in 1927 ultimately based on quantum mechanics. (Very soon after Schrodinger had shown the equivalence of the wave and matrix versions of quantum mechanics in 1926, the unification of these two versions of quantum mechanics was then presented in the "transformation theory" by Dirac and Jordan. Heisenberg thought of and thought over the uncertainty principle when he was studying this "transformation theory" in 1927, to be exact.) Specifically, when Heisenberg was puzzling over the basic quantum properties of electrons, he realized that the act of measuring an electron's properties by hitting it with gamma rays would alter the electron's behavior—that is, the act of observing alters the reality being observed. For instance, to measure the properties of a particle such as an electron, one needs to use a measuring device, usually light or radiation. But the energy in this radiation affects the particle being observed. So he had a shocking but clear realization about the limits of physical knowledge. And his historic realization led to the birth of the famous uncertainty principle!

The historical fact above has clearly told us: the origin of the uncertainty principle is quantum mechanics; that is, quantum mechanics is the

THE CERTAINTY MECHANISM BEHIND THE UNCERTAINTY PRINCIPLE

mother or root of this principle. This origin explicitly and sufficiently demonstrates that the uncertainty principle is about the electrons that emit photons, because the task of quantum mechanics is to describe how electrons radiate or emit photons, as just mentioned above. The major features of this principle are: you can measure the speed of an electron or its position, but not both at once; the better you measure the electron's position, the less accurately you can know its velocity, and vice versa. And if one likes to go a little further, these major features can be described more accurately as: uncertainty, or imprecision always turns up if one tries to measure the position and the momentum of a particle, like an electron, at the same time (momentum is a quantity of motion of a moving object, measured as its mass multiplied by its velocity). For example, an electron cannot be observed without changing its momentum, resulting in the imprecision or uncertainty in the measurement. The similar uncertainty relation also appears when one simultaneously measures the variables of the energy and time of particles, like electrons. Furthermore, according to the interpretation of Heisenberg, these uncertainties or imprecisions in the measurement were not the fault of the experimenter; they were inherent in quantum mechanics instead. (Narrator: such an interpretation, needless to say, is also a further demonstration or noticeable reminder that the origin of the uncertainty principle is quantum mechanics indeed. In addition, the fully recognized, well-known *fact*, which is that the uncertainty principle is the fundamental nature of quantum mechanics, is also a constant and explicit reminder of this origin.)

The uncertainty principle is one of the most significant physical laws, at least being viewed in this way by many physicists. In fact, this principle is prevalently regarded as one of the most fundamental and influential principles in physics, thanks to its multiple functions in interpreting of various quantum phenomena. As a result, being parallel with quantum mechanics and the complementarity principle (which was put forward by Niels Bohr in 1928, primarily for describing the wave/particle duality of light), the uncertainty principle has been widely recognized and accepted as one of the three main fundamental theoretical pillars of quantum physics. With such a great and high status, the uncertainty principle is undoubtedly a wonderful and amazing theory in science!

Wonderful and amazing though, what's equally impressive or unforgettable is that the uncertainty principle is also a theory that has brought about the most intense dispute or disagreement in the history of science (at least in the 20th century). In the paragraph above, we have already seen enough agreeing opinions. To be balanced, objective and fair, we also need to know some representative disagreeing voices over this principle. These disagreeing voices have not only come from many philosophers, but also from some really top scientists. To philosophers, their disagreement was primarily due to their belief that there ought to be an essential or deep reason underlying the uncertainties or imprecisions of the uncertainty principle. Nevertheless, given that this principle is a theory in the field of science, the disagreeing voices from scientists are, of course, more tangible and convincing, more related and important, thus being our main concerns here.

The disagreeing voices from some highly influential scientists can be summed up into two types. One is the most notable, also the most intense, voice. This voice, as many people have been familiar with, was from the great Albert Einstein. His disagreeing voice was clearly reflected in his well-known opinion or statement that "God does not play dice with the universe." And for a long time, Einstein insisted on the uncertainty principle was not a complete theory. Another is the most interesting voice of disagreement. This probably belonged to Erwin Schrodinger, considering he was the very founder of quantum mechanics, whereas quantum mechanics was the origin of the uncertainty principle, as just mentioned above. He disagreed with the indeterministic, statistical viewpoint of the uncertainty principle; and he believed that the statistical interpretation of this principle or its probabilistic implication was due to a yet undiscovered deterministic law. In the camp of Schrodinger, another important figure was Max Planck, the earliest founder of quantum theory (he suggested the quantum hypothesis that initiated the quantum idea in 1900, the year in which the era of modern physics began). Planck had a similar opinion as Schrodinger on the issue of disagreeing with the uncertainty principle.

What appears to be as equally important as those disagreeing voices seems to be that the uncertainty principle has to face the plain *reality* that it couldn't be understood—or at least couldn't be really understood, either in truth or in essence or in both. (For various historical reasons, when the incomprehensible feature of quantum mechanics was involved, more often than not, quantum mechanics and quantum physics were mentioned interchangeably. But the historical background clearly told us that the main target of this incomprehension was the uncertainty principle. In fact, this principle, being the inherent or fundamental nature of quantum mechanics, is frequently said as quantum mechanics in many situations. Anyway, the clear fact is that the uncertainty principle plays a major role in the unfathomable feature of quantum mechanics or quantum physics.)

This plain *reality* is clearly reflected in the two opinions on the uncertainty principle from two famous scientists or physicists. One is probably the frankest opinion, which was from Richard Feynman (he played a crucial role for the birth of quantum field theory, one of the important branches of modern physics. He was a Nobel laureate in physics). His opinion was often quoted as: "I think I can safely say that nobody understands quantum mechanics." And he also frankly stated that he never understood quantum mechanics. (Commentator: the opinion of Feynman openly admitted, or at least clearly indicated, that the uncertainty principle couldn't be understood at all.) Another is perhaps the most confusing or interesting opinion, which came from Niels Bohr. (He was one of the highly influential physicists in the last century. Both he and his son were the Nobel laureates. Specifically, Niels Bohr was one of the strongest supporters of the uncertainty principle, also the master of the Copenhagen interpretation of quantum mechanics.) The opinion of Niels Bohr was shown in his famous statements: "If you are not confused by quantum physics then you have not really understood it." "Anybody who thinks they understand quantum physics is wrong." (Commentator: his opinion at least openly admitted that it is not possible to understand the uncertainty principle, which indicates that even the strong supporters of this principle have to face up to the reality that the uncertainty principle couldn't be understood.) All in all, the plain *reality* that the uncertainty principle couldn't be understood is actually a widely recognized or well acknowledged fact, believe it or not. (Related questions and answers: why do you bring up this plain reality here? Answer: because this plain reality, as readers will see or perceive in this chapter, turns out to be closely related to the intense dispute or disagreeing voices over the uncertainty principle just mentioned above; because this plain reality is known to be one of the most important attributes of the uncertainty principle, also one of the long-term concerned issues about this principle. Questions: is this plain reality closely related to the topics of this chapter? Why? Answer: yes; because the key to knowing the crux or essence of this plain reality lies with revealing the certainty mechanism behind the uncertainty principle.) (As to the long-standing, fundamentally important question: why is it impossible to understand the uncertainty principle? The clear answer to it will appear before the end of this chapter, I promise. So at this moment, let us still continue to keep our attention to the long-term unsolved, intense dispute over the uncertainty principle.)

What does the long-term unsolved, profound and famous, intense dispute over the uncertainty principle really tell us? It tells us that either side seems to need truly reliable evidence for supporting their arguments, and that either side needs the most effective evidence for finally resolving this intense dispute. Then what is the truly reliable evidence? What is the most effective evidence? Or what is the key to finally resolving this intense dispute? Let me unveil a small part of the secret: the truly reliable evidence, also the most effective evidence, thus the key to finally resolving this intense dispute, turns out to lie with the answer to the essentially important question: is there, and what is, the *certainty* mechanism behind the *uncer-*

tainty principle? This answer will be gradually unfolding in this chapter! (Commentator: in the eyes of some people, especially some of those professional people in physics, the intense dispute over the uncertainty principle seems to have become the cloud and mist of history. However, it is a reality, a today's reality, also a clear and inescapable *reality* that reminds, even warns, us: it is too early to claim that this intense dispute has been over; and it seems not wise to put aside such a profound, highly influential, long-term unsettled, intense dispute. In fact, this *reality* is not only closely related to this intense dispute, but also far more profound and important than this intense dispute itself, because this *reality* has crucial implications or fateful consequences.) Then what reality is so profound? What reality is so important? What reality is so crucial or fateful? It will appear immediately!

This clear and inescapable *reality* is: quantum mechanics and general relativity cannot both be correct because they are known to be inconsistent with each other, as explicitly pointed out by Hawking in his *A Brief History of Time* (1998 version, P. 12). (Related questions and answers: what does such a *reality* really tell us? Or what are the profound or true consequences of this *reality?* Answer: the apparently glorious mansion of modern physics turns out to be seriously shaky, or at least cannot be stable, being the unavoidable, also extremely serious, consequences of this *reality*. This answer, sounding somewhat "harsh" though, is clear, even obvious, because quantum mechanics and general relativity are the two main theoretical pillars of modern physics. In fact, it is quite safe and fair if one concludes that all today's qualified or insightful experts in physics, especially those in theoretical physics, might have already realized the terrible aftermath of the long-standing, serious inconsistency of quantum mechanics and general relativity, because they are well able to be aware that this serious inconsistency is not only the greatest challenge in theoretical physics, but also the deepest disaster of modern physics. More than that, when viewed from a wider angle, this serious inconsistency has also been openly admitted as one of the most important, long-term unsolved problems in science. One may then ask: what is the relation between this *reality* and the uncertainty principle? Answer: the uncertainty principle is the fundamental nature of quantum mechanics, being a well-known fact in modern physics as a clear fact, as a fully recognized fact, also as an irrefutable or undeniable truth; along with that quantum mechanics is the origin of this principle, as just pointed out above; and the present interpretation of this principle says that there is no certainty mechanism behind it. So with these three explicit reasons, the uncertainty principle, especially via the present interpretation of this principle, can be inherently or closely related to this *reality*.) Accordingly, in the face of such a clear and inescapable *reality*, and in

THE CERTAINTY MECHANISM BEHIND THE UNCERTAINTY PRINCIPLE

front of the unavoidable, extremely serious consequences of this *reality*, it becomes rationally clear or sensible that we cannot afford not to ask the crucially important question with profound and far-reaching implications, and with great importance or significance: is there, and what is, the *certainty* mechanism behind the *uncertainty* principle?

With this crucially important question as a specific and marked reminder, the central task of this chapter, which is about briefly introducing how to reveal and determine the *certainty* mechanism behind the famous *uncertainty* principle, thus becomes essentially more profound, fundamentally more important and scientifically more meaningful (than it was first mentioned above). Moreover, since the uncertainty principle is (said to be) a wonderful and amazing theory, clearly, even obviously, also plainly and undoubtedly, the *certainty* mechanism behind this principle will be even more wonderful and amazing. In other words, this central task thereby becomes much more wonderful and amazing (than originally thought). This wonderful and amazing certainty mechanism is becoming closer and closer to us. We are gradually, step by step, approaching this wonderful and amazing certainty mechanism! The central task of this chapter, or the *certainty* mechanism behind the *uncertainty* principle, is about to spread out its wonderful and amazing charm to all of us!

(With such an attractive certainty mechanism waving its warm greetings to us, some readers may eagerly, even hastily, look forward to arriving at the keenly expected destination of the journey of this chapter: they want to witness the fascinating, wonderful certainty mechanism behind this principle as soon as possible, as quickly as possible. Yes, the spectacular scenery on the top of this destination can certainly make people excitedly yearn for, but surely enough, the marvelous views along the route towards this destination will also be touching your heart and soul with their amazing and superb wonders. Moreover, after experiencing these superb wonders, you may cherish and enjoy the spectacular scenery on the top of this destination even more: this scenery appears much more spectacular and precious to you, after your heavy sweating over climbing these amazing wonders on the route. So, in order to make sure dear readers can really enjoy this spectacular scenery as much as possible, let us start our journey from the amazing and superb wonders along the route to our final destination—the *certainty* mechanism behind the *uncertainty* principle. No worry: I am your guide in this journey; perhaps, I am the person who knows this route best in the world. Of course, whether I am a competent guide in the coming journey, dear readers will have a final say after this journey.)

But in order to carry out the central task of this chapter, it is necessary first to introduce the newly discovered law that has eventually solved the problem of *why* there are quantum states, by revealing the *quantum* mechanism of *why* and how photons, being the tiny and discrete quanta or particles of light, get their velocity c (the speed of light) from the electron that emits them. (Related question and answer: why are you going to introduce such a new law before carrying out this central task? Answer: as mentioned earlier in this book, quantum mechanics is unable to solve the most fundamental problem in front of itself, *why* there are quantum states, simply because it doesn't have the ability to reveal the mechanism behind this *why* at all—further specifically thus more perceptibly, because quantum mechanics doesn't have the ability to reveal the *quantum* mechanism of why and how photons get their velocity c (the speed of light) from the electron that emits them. What should be mentioned is that this inability of quantum mechanics has been perplexing many brilliant physicists for a long time; that is, this inability, being a tacitly admitted *fact* in reality, has become an actually recognized or clearly realized *fact* in essence, thus also being an irrefutable or undeniable *fact* in truth. More specifically to be even more noticeable and impressive, as pointed out earlier in this book, one could also clearly perceive and explicitly realize, even merely from the large perspective or basic features of quantum mechanics, this inability of quantum mechanics, if he or she views it or thinks it over conversely as follows. If quantum mechanics had been able to solve this most fundamental problem, its crucial or core *assumption*, also the fundamentally indispensable *assumption* for the birth of quantum mechanics, which says that all forms of energy released from electrons are in the way of discrete units or bundles called quanta (which are known as photons nowadays), would be utterly unnecessary; thus this core *assumption* would never have appeared at all. And so, the irrefutable and unavoidable *fact* that quantum mechanics indispensably and desperately necessitates this core *assumption* can enable one to perceive this inability of quantum mechanics clearly and realize it explicitly; this irrefutable and unavoidable *fact* is also a clear demonstration or impartial witness that one cannot deny this inability of quantum mechanics by any means; this irrefutable and unavoidable *fact* is, of course, also a clearly seen fact to the related professional people in physics. All in all, this inability of quantum mechanics is not only a clear *fact*, but also an undoubted or undeniable *fact*.) What should be pointed out or noticed is that this inability of quantum mechanics, as readers will see in this chapter, turns out to be the deep reason why quantum mechanics is unable to reveal the *certainty* mechanism behind the *uncertainty* principle. By contrast, this new law, as will also be seen in this chapter, because it has solved the problem of *why* there are quantum states, is indispensable for revealing the certainty mechanism behind this principle. So a brief in-

THE CERTAINTY MECHANISM BEHIND THE UNCERTAINTY PRINCIPLE

troduction or explanation of this new law is the necessary first step to accomplish the central task of this chapter.

At this moment, some dear readers may still be curious: what is the original purpose of this new law? I cannot bear not to answer this kind of curiosity. The feature of photons (the tiny and discrete quanta or particles of light), which is the particle nature of light, was gradually known in the early 20th century. Initially, in 1900 this feature was reflected in Planck's quantum theory for interpreting the energy radiated from a hot object or body, like an oven or a star. Shortly after that, in 1905, this feature was employed in Einstein's interpretation of the photoelectric effect, the phenomenon of ejecting electrons from a metal shone by high-energy light. Then, in 1923, again this feature was demonstrated in the famous Compton scattering experiment, in which the high-energy rays of light, X-rays, were used to strike electrons. Through these remarkable achievements (all of them were awarded the Nobel Prize), the feature of photons has been accepted; and it has also been known that photons are emitted from electrons. (In fact, virtually all the photons experienced in our daily life are emitted from electrons. There are quite a few such examples: the photons from family light bulbs, the photons from portable flashlights, and the photons from vehicle lights, even the photons from the open fire of your barbecue dinner, and so forth. Moreover, some of the most important, also the most influential, theories in modern physics, established in the early 20th century, have also fully recognized the plain actuality that photons are emitted from electrons. For instance, the well-known quantum mechanics, developed by Schrodinger and Heisenberg in the 1920s, openly admits that photons are emitted from electrons; and the famous model of the hydrogen atom, put forward by Niels Bohr in 1913, clearly and explicitly tells us that photons are emitted from electrons. What should be mentioned or pointed out is: both these theories are the crucial materials in the textbooks on modern physics; that is, it is rather rational and quite safe to say that the professional people in physics must have known these two theories very well. In other words, it is both rational and realistic if one concludes that all the qualified or knowledgeable professionals in physics are very familiar with this plain actuality.)

However (this is a heavy HOWEVER with profound implications!), the *quantum* mechanism of why and how photons get their velocity c (the speed of light) from the electron that emits them still remains a big mystery. This is clearly reflected in the two pieces of hard evidence coming from two historical facts. One fact is that Einstein had to *assume* "The photon remains localized in space as it moves away from the source with a velocity c" (again c is the speed of light), for his interpretation of the photoelectric effect in 1905. (Narrator: if Einstein had known this *quantum* mechanism, he would not have employed such an assumption at all, simply because it was utterly unnecessary. In other words, the absolute necessity of employing this very assumption in his interpretation of the photoelectric effect was actually an explicit demonstration or clear indication that this *quantum* mechanism was unknown at that time.) Another fact is also from Einstein: "All these fifty years of pondering have not brought me any closer to answering the question, what are light quanta?" (Commentator: if Einstein had known the *quantum* mechanism of why and how photons get their velocity c, the speed of light, from the electron that emits them, he would have clearly known "What are light quanta?")

What should be pointed out is: clearly, even obviously, also self-evidently and indisputably, the problem of *why* and *how* photons, being the tiny and discrete quanta or particles of light, get their velocity c (the speed of light) from the electron that emits them (that is, the problem of *why* there are quantum states) is the most fundamental problem in the realm of entire quantum physics, also the most crucial and most important problem in this realm. Accordingly, also quite obviously and definitely, solving this problem is not only fundamentally important, but also crucially or remarkably significant. Therefore, the original purpose of this new law is for revealing the *quantum* mechanism of why and how photons get their velocity c, the speed of light, from the electron that emits them—that is, for solving the long-term unsolved, fundamentally important problem: *why* there are quantum states. In addition, a related clarification is provided here as follows. The prevalent idea, which treats photons as force carriers for transmitting electromagnetic force, though highly influential at present, still doesn't and can't solve the problem of why and how photons get their velocity c (the speed of light) from the electron that emits them; not to mention that this idea has never targeted at solving this problem at all.

Having previewed the background information about the necessity of introducing the newly discovered law, let us go to it now. This new law of physics came to the world recently (discovered just a few years ago by me, the author of this book[*]). ([*]But for any law or theory in science, including any law or theory in physics of course, a well-known, also fully recognized and totally accepted, basic principle is: the things that really matter lie in *what* rather than who, lie with *what* the law or theory deals with or talks about, instead of who found or developed it. In fact, it is not only definitely rational but also surely reasonable if one concludes that all today's qualified or knowledgeable scientists must have clearly known and totally agreed with such a basic principle, because it is the cornerstone of science or can play the role pretty similar to a cornerstone; because it has been completely recognized and accepted by scientific community as general knowledge or common sense in science nowadays. That is, one can safely say that no experienced or eligible scientists would have objection to this basic principle. Moreover, without this

THE CERTAINTY MECHANISM BEHIND THE UNCERTAINTY PRINCIPLE

basic principle or if this basic principle were thrown away, science would lose one of the rational and objective criteria; the most fundamental nature of science would be fatally damaged; the most fundamental spirit of science would no longer exist; science would no longer be science at all! Therefore, it is actually, or should be, not only rather rational but also quite reasonable if one comes to such a simple and clear conclusion: all today's qualified or experienced scientists would actively persist in, conscientiously uphold, and painstakingly defend this basic principle. Correspondingly, in order to ensure the integrity of science through keeping its rationality, objectivity, truthfulness and reliability, there is also such a very basic professional requirement for anyone to introduce a scientific law or theory—no matter who discovered or developed it: he or she must NOT make any exaggerated descriptions about the law or theory, including its meanings, implications and significance. More explicitly to be further clear and definite, he or she must only tell truth and fact; he/she must ensure that only truth and fact are introduced or stated. The response or promise from me: surely and of course, I will strictly obey this very basic professional requirement in introducing this new law!) Having said those, let us enter this new physical law from the door of its most important part, also its most fundamental part, which is its dynamic elements or components; that is, what forces are involved in this law.

This law includes two forces with opposite directions (Fig. 6.1, next page). One is the newly discovered electron-self-exerting magnetic force, which is the magnetic force that acts on an orbiting electron through the magnetic field generated by the orbiting motion of the electron itself. (When the electron orbits clockwise, the direction of this magnetic field is out of paper, or towards you, *determined by the opposite orbiting direction of an equivalent positive charge because an electron has a negative charge;* but its direction is into paper, or away from you, when the electron orbits counterclockwise, *still determined by the opposite orbiting direction of an equivalent positive charge,* based on the well-known, right-hand rule that is applicable to the situation of the electric current in a loop wire. This rule is available in the textbooks for middle or high schools. So the existence of this magnetic field has pretty solid theoretical grounds; thus the discovery of this magnetic field should or could be easily acceptable to most people.) The magnitude or value of electron-self-exerting magnetic force depends on three factors: the charge of an electron, its orbiting velocity, and the strength of the magnetic field generated due to the orbiting motion of the electron. The direction of this force is radially outward from the nucleus orbited by the electron to this electron, which is determined by the famous right-hand rule in electromagnetism *via the opposite orbiting direction of an equivalent positive charge* (what should be noticed is that the direction of this force is *always* radially outward from the nucleus to the orbiting electron, no matter whether it orbits clockwise or counterclockwise, because the magnetic field is generated by this orbiting electron itself). This newly discovered force is crucially important to the dynamics of this new law. Another is electric force acting on an electron; that is, Coulomb force, which is familiar to many people (because this force is a highly important force that appears in the textbooks written for middle or high schools). The magnitude or value of electric force is determined by two factors: the charge of an electron and the strength of the electric field where the electron resides; the direction of electric force is radially inward the nucleus orbited by the electron from this electron. These two forces together determine the dynamics of this new law.

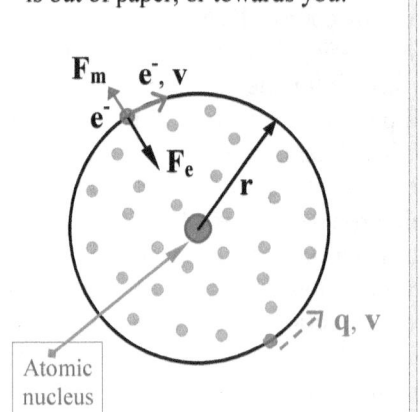
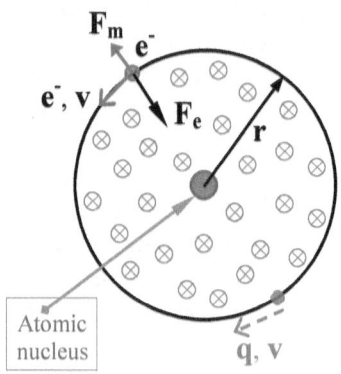

e^- — an orbiting electron,
v — the orbiting velocity of the electron,
r — the radius of the position of the equilibrium orbit of the electron,
F_m — electron-self-exerting magnetic force,
F_e — electric force or Coulomb force,
q — the equivalent positive charge.

Figure 6.1, the two forces that simultaneously act on an orbiting electron.

Since the concept of electron-self-exerting magnetic force is crucially important in this new law, it is fundamentally important to grasp this concept effectively if one wants to enter this law quickly. Considering that the appearance of this force may be a surprise to some people (because it was never discovered before the birth of this law), it seems better that there ought to be a good emphasis on how to perceive and realize the concrete and definite meanings of this concept.

The key to grasping this concept lies in explicitly sensing that electron-self-exerting magnetic force turns out to be an objectively existing physical force. This is not difficult because there are sufficient, irrefutable theoretical bases and a large number of daily-seen practical experiences or facts to support, show and substantiate that. First, the existence of electron-self-exerting magnetic force has quite solid theoretical grounds. The related theories in physics clearly and unavoidably indicate: since an orbiting electron inevitably generates a magnetic field—a universally accepted fact, why doesn't this magnetic field exert a magnetic force on the orbiting electron itself? With this universally accepted fact as a specific and noticeable reminder, it seems not difficult that one can easily realize: it turns out that electron-self-exerting magnetic force comes from the fundamental principles of electromagnetism—specifically, from the basic rules on how a moving charge generates a magnetic field and how a charged particle moves in the magnetic field (these basic rules, because they are available in the textbooks for middle or high schools, are familiar to today's educated people). With this realization, it seems pretty reasonable and fair that one can clearly and definitely perceive: any denial of the objective existence of this newly discovered magnetic force turns out to be directly and flatly contradictory with the fundamental principles and basic rules in electromagnetism. Quite obviously, also rather rationally, such an explicit perception is substantially helpful for one to grasp the concept of electron-self-exerting magnetic force effectively, clearly and easily.

Second, the existence of electron-self-exerting magnetic force also has quite solid practical grounds, being impartially witnessed and explicitly substantiated by a host of obviously valid daily experiences or daily-seen facts. For instance, without the action of this force, electrons would be dragged onto atomic nuclei in the blink of an eye (dynamically like meteorites striking onto the earth) once we switch on the lights, then our family light bulbs would not work; ordinary metal electrical conductors would not work; various and numerous electrical equipment would not work. In other words, if the action of electron-self-exerting magnetic force were taken away, our daily experienced world would, in fact, become literally inconceivable. So the numerous, related, daily-seen practical experiences or facts in our modern life have no choice but to show us that the newly discovered electron-self-exerting magnetic force is indeed an objectively existing physical force! In addition, if one thinks deeply, he or she can suddenly, but clearly and surely, realize: without the action of electron-self-exerting magnetic force, there would be no grounds to explain why photons—being the tiny and discrete quanta or particles of light, either emitted from 3-volt portable flashlights, or from 12-volt vehicle lights, or from 110-volt or 220-volt family bulbs, have the same velocity c (the speed of light). This realization can also enable one to perceive that really and surely, electron-self-exerting magnetic force turns out to be an objectively existing physical force.

Having grasped the concept of electron-self-exerting magnetic force by realizing that this force turns out to be an objectively existing physical force, we get prepared to look at the main dynamic features of this new physical law. On the whole, the dynamics of this law describes the orbiting motion of an electron under the simultaneous action of the two forces with opposite directions (Fig. 6.1, P. 170). One is electric force with its direction radially inward the nucleus orbited by the electron from this electron. Another is electron-self-exerting magnetic force with its direction radially outward from the nucleus orbited by the electron to this electron. So it is the resultant force of these two forces that provides the centripetal force for an orbiting electron (centripetal force is a force which makes things move towards the center of something when they are moving quickly around that center).

Specifically, under the joint action of these two forces, the entire dynamic process consists of several sections or parts (Fig. 6.2). (1) The effect of exerting an external electric potential difference is reflected in its increasing the orbiting velocity of an electron, based on the work done on the electron equal to its kinetic energy change, one of the most fundamental principles in physics. (Even the external electric potential difference is not high—such as 3 volts, the electron's orbiting velocity is still increased rapidly. This is a bit like the situation in which a charged particle is still speeded up very quickly even in a comparatively weak electric field.) (2) Continuously doing work on the orbiting electron under the action of the external electric potential difference results in a quick increase in the orbiting velocity of the electron, thus a rapid decrease in its orbiting radius (the larger the orbiting velocity of an electron, the smaller the orbiting radius of the electron). Corresponding to this increase and decrease, while both electric force and electron-self-exerting magnetic force increase quickly, the latter increases much more rapidly than the former (because the latter is increased by two factors: the increase in orbiting velocity and the decrease in orbiting radius, whereas the former is increased only by one factor: the decrease in orbiting radius). (3) This crucial difference in their increasing rates determines that the latter will catch up with the former very rapidly, thus reaching a special moment at which these two forces are equal to each other in value. At this special moment, the instantaneous maximum velocity of the electron is equal to the speed of light c (this is determined by the core dynamic equation of this new law). And at this special moment, a photon is emitted from the electron. (Note: at the moment at which the instantaneous maximum velocity of an electron is equal to the speed of light c, the mass of the electron is instantly at zero, based on the mass consumption caused by mass doing positive work. About the mass consumption, please see chapter two if necessary.)

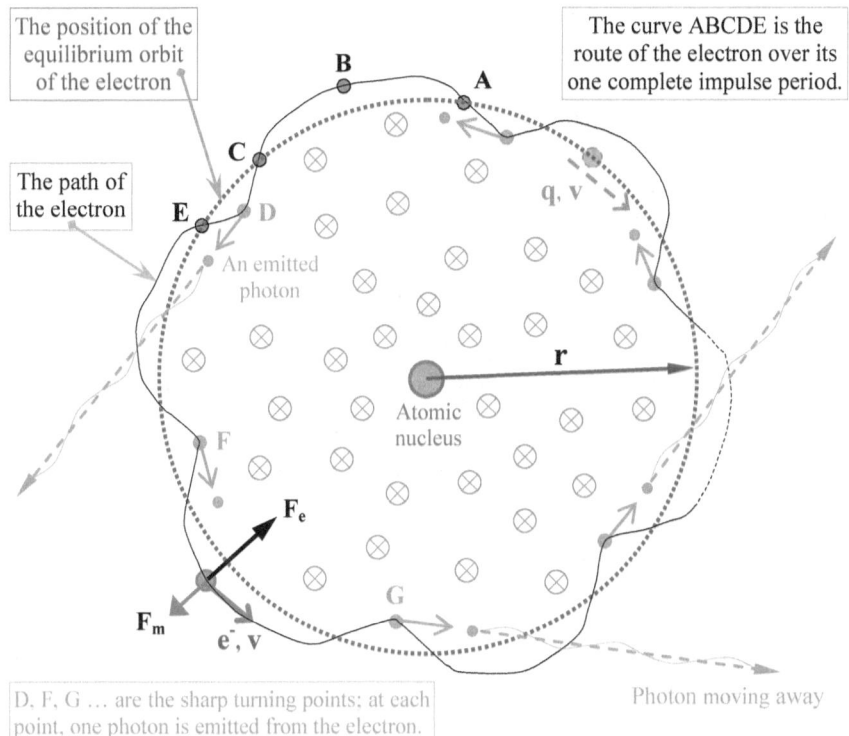

⊗ **The direction of the magnetic field is into paper, or away from you.** ⊗

e^- — an *orbiting* and *periodic impulses* electron,	F_m — electron-self-exerting magnetic force,
v — the orbiting velocity of the electron (which is the velocity along the direction of the position of the equilibrium orbit of the electron),	r — the radius of the position of the equilibrium orbit of the electron,
F_e — electric force or Coulomb force.	q — the equivalent positive charge.

Figure 6.2, the law of an orbiting electron with periodic impulses emitting photons (a newly discovered physical law). This new physical law (for the first time in the history of science) reveals the mechanism of *why* and *how* photons, being the tiny and discrete particles of light, get their velocity c (the speed of light) from the electron that emits them, solves the fundamental problem of *why* there are quantum states, and unifies the wave/particle dual natures of light.

With these main dynamic features, the newly discovered physical law is unfolding its most fundamental role to us through the key features as follows. Overall, this law dynamically unveils the *mechanism* of why and how the entire motion of an electron that emits photons consists of two simultaneously existing movements: orbiting and periodic impulses closely around the position of its equilibrium orbit (Fig. 6.2). That is, an orbiting electron simultaneously undergoes periodic impulses—alternately and closely passing inside and outside the position of its equilibrium orbit. Specifically, this law dynamically reveals that over one complete impulse period of an electron, there is one moment at which the instantaneous maximum velocity of the electron is equal to the speed of light c at the sharp turning point on its path closely inside the position of its equilibrium orbit; and at this point, a photon is emitted from the electron (this means that over one complete impulse period of an electron, only one photon is emitted from the electron). As a result, all the photons emitted at these sharp turning points inherit the instantaneous maximum velocity c from the electron that emits them, maintaining and moving away at velocity c (the speed of light). Therefore, this law has eventually solved the fundamental problem of *why* there are quantum states, by revealing the *quantum* mechanism of *why* and *how* photons, being the tiny and discrete quanta or particles of light, get their velocity c—the speed of light, from the electrons emitting them. (What should be pointed out or noticed is: this law, and only this law, has the ability to solve this fundamental problem.)

Due to the key features presented above, this law has been named (by me) as the law of an orbiting electron with periodic impulses emitting photons (OEPIEP law, for short). (Related questions and answers: what is an easy or tangible way to think of OEPIEP law? Answer: when we switch on an electric light, it gives off photons; when we switch off an electric light, it does not. And the question "Where does the velocity of photons come from?" can enable one to think of OEPIEP law easily and tangibly, because this law is the *first* and *only* physical law or theory, *in the history of science*, that has answered this question, as just mentioned above. Moreover, this question and the answer to it open the door of OEPIEP law, and lead one to enter this law directly. Question: a fundamentally important and definitely necessary question to the newly discovered OEPIEP law— has OEPIEP law been verified or confirmed? Answer: yes, it has been; in fact, it has been comprehensively verified or confirmed from different aspects; about this crucially or extremely important issue, it will be mentioned subsequently in the relevant parts. So at this moment, it is enough to know in advance that these verifications or confirmations have clearly shown that the results obtained from OEPIEP law agree with experiments and observations very well; thus the validity and reliability of OEPIEP law are quite positive. Question: besides the key features described above, does OEPIEP law have profound and wonderful roles or meanings? Answer: yes, it has; and these roles are reflected in the several paragraphs to be coming.)

This new physical law does have profound and wonderful roles! These roles are clearly displayed in the three great tasks it has fulfilled. Because these tasks are crucially important, let us succinctly go over them one by one. First, this law has revealed the connection in velocity between an electron and its emitting photons by showing why and how photons get their velocity c (the speed of light) from the electron that emits them. This revealing eventually answers the long-standing question: what is the fundamental connection between an electron and its emitting photons? This answer is of fundamental significance, because one of the most important purposes of science is to discover the inherent connection of different, but intrinsically related, phenomena. (Commentator: clearly, also inescapably, an important implication of this role is that no earlier theories, including quantum mechanics, were able to reveal this fundamental connection!)

Second, this law has revealed the mechanism behind Einstein's famous assumption that "The photon remains localized in space as it moves away from the source with a velocity c." (Note: c is the speed of light. This assumption, due to its crucial function in explaining the photoelectric effect—the phenomenon of ejecting electrons from a metal shone by high-energy light, and because of the important status of this phenomenon in physics, is known to be one of the most important and influential assumptions in modern physics. In fact, one can safely say that all the professional people in physics must/should have known this assumption very well.) This revealing would probably make Einstein feel relieved if he were still alive today, because it ultimately answers his unfathomable enigma of "What are light quanta?" More than that, the profound implication of this revealing lies with: it turns out that there is the solid mechanism behind this important assumption. (Narrator: clearly, also unavoidably, what accompanies this role is the regret of having not revealed the mechanism behind this famous assumption in the past.)

Third, this law has at last injected the true meaning into the remarkably important phenomenon in physics, electromagnetic radiation—that is, electrons emit photons. (As mentioned earlier, this phenomenon had played a crucially important role at the critical turning point in the development from classical physics to modern physics. Its crucially historic significance is thereby displayed.) In this law, the dynamics of electro*magnetic* radiation includes both electric force and *magnetic* force. On the contrary, prior to this law, no dynamics of electro*magnetic* radiation included *magnetic* force at all. For instance, neither the famous model of the hydrogen atom nor the well-known quantum mechanics included magnetic force—that is to say, the action of *magnetic* force utterly evaporated in their descriptions of electro*magnetic* radiation; consequently, the so-called electromagnetic radiation turned out to be merely an empty terminology in truth! So, in terms of describing electromagnetic radiation, there is a big difference before and after this law. And through this big difference, actually a day and night difference, one can clearly see the serious flaw in the related "traditional" theories. (Commentator: as a result, the birth of this law eventually ends the era, in which the description of the phenomenon of electro*magnetic* radiation did not include the action of *magnetic* force, and begins the era in which the description of electro*magnetic* radiation includes the action of *magnetic* force! And so, the fundamentally important phenomenon of electro*magnetic* radiation begins to have or be given the description that is really worthy of its name from now on. So congratulations go to you, electro*magnetic* radiation—the fundamentally and crucially important phenomenon in physics!)

Yet what appears even more profound and wonderful (than the three profound and wonderful roles just mentioned above) is perhaps the unique feature of this new law (that is, the law of an orbiting electron with periodic impulses emitting photons). This unique feature lies with: the core dynamic equation of this law is the first equation, in the history of science, that *simultaneously* possesses wave nature and particle nature. So, this feature dynamically reveals that an orbiting electron, a discrete tiny particle though, possesses wave nature at the same time—its wave nature is explicitly reflected in its periodic impulses closely around the position of its equilibrium orbit (Fig. 6.2, P. 173). Moreover, this feature shows that the wave/particle dual natures of the electron are dynamically inseparable, thus inherently unified. As a result, photons (the tiny, discrete particles of light) not only inherit the wave/particle dual natures from the electrons emitting them, but also maintain the unification of their inheriting dual natures. (Note: the moving pattern of photons, after they are emitted from electrons, cannot be changed by either electric force or magnetic force or any of the known forces.) Therefore, this new physical law, through the unique feature above, unifies the wave/particle dual natures of light by dynamically unifying the wave/particle dual natures of the electrons emitting light. This unification, and only this unification, can demonstrate that the wave nature and particle nature of light turn out to be fundamentally compatible thus intrinsically consistent with each other! In fact, this unification literally shows that the wave nature and particle nature of light turn out to be inherently <u>connected together</u>. (Related question and answer: how could one see this unification more clearly thus realize it more explicitly? Answer: the information in the coming paragraph can be very helpful.)

THE CERTAINTY MECHANISM BEHIND THE UNCERTAINTY PRINCIPLE

This unification can be seen more clearly through the closely related questions and answers like the following. Question: what is the key to understanding this unification? Answer: after revealing the mechanism behind the phenomenon of the wave/particle dual natures of light, its dual natures turn out to be inherently unified from their roots. So, this unification is also a clear indication that in science, it is of crucial importance to reveal the mechanism behind an observed phenomenon. Question: what is the characteristic feature of this unification? Answer: it is the unification between the observed phenomenon, which is the wave/particle dual natures of light, and the mechanism causing this phenomenon. Thus, this unification is also an in-depth unification. Question: are there available experiments that are consistent with this in-depth unification? Or does this in-depth unification have solid experimental grounds? Answer: yes, there are; yes, it has. These experiments include the appearance of single-slit diffraction and double-slit interference, when *only* a single photon at a time is emitted from the source of light. In fact, if one spends a few minutes thinking over deeply, he or she can clearly realize: only the unification of the wave/particle dual natures of light can truly explain these two kinds of experiments. Questions: does this in-depth unification have profound implications with fundamentally important or far-reaching significance? If yes, what are they? Answer: the pieces of information in the coming several paragraphs will give you the answers.

One of the profound implications lies with that this in-depth unification eventually unravels the long-standing famous puzzle: why light, while being electromagnetic *waves*, does not require any medium for its propagation, such as the light from the sun to the earth. To be clarified, the wave nature of light *on its own* is utterly unable to explain the fact that light can travel through totally empty space, because this kind of space does not have recognizable physical medium for propagating light *waves*, whereas the *wave* nature of light *on its own* must require a medium for its propagation, like sound waves going through air. (This point could be clearly reminded by the well-known historical event about looking for the "ether" that occurred in the second half of the 19th century. The claimed "ether" was then believed as the medium propagating light waves.) Now, it is clear: only this unification can really explain the known fact that light, electromagnetic *waves* though, can travel through totally empty space. (Without this unification, one of the most ridiculous or awkward situations in the world could appear as follows. According to its wave nature, light could not travel through totally empty space, whereas according to its particle nature, light could. Then if one asks light: "Can you travel through totally empty space?" "I don't know," or "I can't know," is probably, and should be, the rational or sensible answer from light. Of course, a ray of light from the sun to the earth cannot tell us: "Only my particle nature comes, but my wave nature still remains on the sun.") All in all, the pieces of concrete information above collectively and clearly point to that the unraveling of this long-standing famous puzzle is indeed a profound implication with fundamentally important or far-reaching significance.

In fact, the real significance of this profound implication, when viewed from the large perspective of the step-by-step development about our humanity exploring the nature of light, will become much more fundamentally important and far-reaching than what has been mentioned above. In other words, this profound implication, when it is considered from the angle of the specific process in mankind's understanding of the nature of light, is actually a remarkable and historic event. Why? Let the facts tell us! Let us see the facts!

Historically, altogether three development stages have experienced about the nature of light. (1) The stage of particle nature. In the late 17th century, Isaac Newton suggested that light was made of tiny particles and our eyes were sensitive to these particles, based on the observation that sunlight could be separated into light with different colors. By then the opinion of the particle nature of light was dominant, partly due to Newton's highly influential status in science. (2) The stage of wave nature. At the beginning of the 19th century, Thomas Young observed that light behaved more like sound waves or water waves, especially when passing through one or more narrow slits (the experimental grounds for wave nature). Then in the 1860s, James Clerk Maxwell established the four famous equations unifying electricity and magnetism as electromagnetism, and showed how light and magnetism were really the two aspects of electromagnetism; that is, light was electromagnetic waves (the theoretical grounds for wave nature). Up to that time, especially through the combination of the experimental and theoretical grounds for wave nature, the viewpoint of the wave nature of light became dominant, whereas Newton's view of particle nature was thought to be wrong. (3) The stage of wave/particle dual natures. In the early 20th century, the particle nature of light (that is, the feature of photons or discrete light quanta) was gradually noticed and revealed (as mentioned earlier, this feature was shown in Planck's quantum theory for interpreting the energy radiated from a hot object or body—such as an oven or a star, in Einstein's interpretation of the photoelectric effect, and in the famous Compton scattering experiment). Moreover, light quanta (the tiny and discrete particles of light) came across as waves when an experiment was designed to observe the wave nature of light; and light quanta came out as particles when an experiment was designed to look for the particle nature of light. With these experimental observations, particularly with the birth of the complementarity principle (which was put forward by Niels Bohr in 1928, primarily for describing or dealing with the wave/particle duality of light), the wave/particle dual natures of light have been widely accepted since then on. To make a long story short, the three development stages about the nature of light are: particle nature → wave nature → wave/particle dual natures.

Yes, the idea of the wave/particle dual natures of light might seem to be content for some people. However, believe it or not (admit it or not), the real or deepest implication of this idea is that the wave/particle dual natures of light are actually *incompatible* with each other. For instance, the wave nature of light *on its own* must need a medium for its propagation, like sound waves going through air, whereas the particle nature of light does not require any medium for its propagation. In other words, according to its wave nature, light must need a medium for its propagation, whereas according to its particle nature, light does not need any medium for its propagation at all. Then if light is asked: "Do you need a medium for your propagation?" "I don't know," or "I can't know," is probably, and should be, the rational or sensible answer from light. Thus, one can clearly see or sense that this actual incompatibility really has serious flaws or inescapable consequences: it is, in essence, unable to explain some fundamentally important phenomena, or at least it cannot give consistent explanations to these phenomena; it is, in fact, incapable of providing a satisfactory explanation for these phenomena.

The real implication of these serious flaws candidly points to: it is fundamentally important to solve this actual incompatibility; the way to do this, and the only way to do this, lies with the unification of the wave/particle dual natures of light. This new physical law (that is, the law of an orbiting electron with periodic impulses emitting photons), through its unique feature presented above, has accomplished the unification of the wave/particle dual natures of light by dynamically unifying the wave/particle dual natures of the electrons emitting light. So with this unification, the entire process of mankind's understanding of the nature of light has altogether undergone four development stages: particle nature → wave nature → wave/particle dual natures → the unification of wave/particle dual natures. As a result—as an explicit and noticeable result, this unification, when viewed from the angle of the specific advancement process in our humanity exploring the nature of light, is indeed a remarkable event with profound implications and with far-reaching, historic significance. Many people, especially those insightful and brilliant scientists who have noticed the serious flaws or inescapable consequences of this actual incompatibility, are probably eagerly looking forward to or waiting for this unprecedented and historic unification! (In addition, this unprecedented and historic unification is also a concrete demonstration that the development of science is a step-by-step process, which, in turn, is a clear indication that the development or advancement of science requires the unceasing efforts from generation to generation. So this indication also clearly and definitely tells the entire world: though the unification of the wave/particle dual natures of light has been eventually completed by me, the great achievements from the diligent, unceasing efforts of former generations are extremely important and indispensable to this profound, far-reaching and historic unification. Therefore, these great achievements should be fully appreciated and highly respected, and ought to be kept firmly in mind forever!)

Up to here, we have experienced the marvelous views brought to us by the newly discovered physical law. With these marvelous views full of the route to our planned destination, which again is the *certainty* mechanism behind the *uncertainty* principle, dear readers may have two typical reactions. Some may feel these fantastic views are so attractive that they tend to forget our destination; others may believe that the spectacular scenery on the top of our destination will be much more attractive, so they are inclined to look forward to arriving at this expected destination even more eagerly and hastily. Either reaction seems to need a friendly reminder: now, we are not far from this spectacular destination; after crossing over a beautiful bridge, we will see it! (Narrator: this bridge, being the most essential connection between an electron and its emitting photons, is electron-photon work-energy relationship or equation. This relationship is the direct result of an application of the new physical law, which is the law of an orbiting electron with periodic impulses emitting photons that you've just read above.)

After visiting the amazing views coming from the new physical law, a beautiful, grand bridge is right in front of us. This grand bridge is the only way to our spectacular destination—the *certainty* mechanism behind the *uncertainty* principle. Then what is this grand bridge? It is electron-photon work-energy relationship (equation)! This relationship shows that, *the amount of __work__ accepted by an electron over its one impulse period is equal to the __energy__ of the photon emitted by the electron (in this impulse period)*. In the face of the imposing grandeur and startling simplicity of this grand bridge, most likely some, even many, readers can't help being astonished: where did this grand bridge come from? How was this grand bridge built?

Most succinctly, electron-photon work-energy relationship is the combination of the new physical law (that is, the law of an orbiting electron with periodic impulses emitting photons, which has just been briefly introduced above) and the famous Planck-Einstein equation (this equation is $E = hf$, where h is Planck constant, f is the frequency of a photon, and E is the energy of the photon). Most plainly, this relationship is simply the result of an application of this new law. (Perhaps, most readers might feel that these answers are too short to allow them to view and appreciate the wonderful fascinations of this relationship. So it seems better to go a little further and deeper.)

Roughly speaking, the idea of electron-photon work-energy relationship initially came from the inspiration of a couple of simple instances in our daily life. One example is: when we turn on a light bulb, photons are emitted (from the electrons in its filament); when we turn off a light bulb, no photons are emitted. However, it has been known that an external electric potential difference (that is, electric power) can do work on electrons but cannot on photons. Thus, here is the puzzle: where does the energy of photons come from? This instance can therefore lead one to think of: if the work accepted by an electron becomes the energy of the photons emitted by the electron, then this puzzle is solved. Another example is: when we switch on a light bulb, what we input is electrical energy, whereas what is output or released is photon energy. This experience may motivate one to ask the question: what is the mechanism that converts the input electrical energy into the released photon energy? So this question can direct one to think over: if the work accepted by an electron becomes the energy of the photons emitted by the electron, this question is then answered. Thus, either of these two instances can easily guide us to an implication or suggestion (at least a clue) that it is very likely that there ought to be the work-energy relationship between an electron and its emitting photons, and this relationship can be essentially important. (This kind of conjecture, being merely a rough or very basic sketch roughly indicating the estimated direction in which one ought to work on, is, of course, still far from showing the existence of the work-energy relationship between an electron and its emitting photons. In order to show its existence, we must find and establish this relationship!)

Overall, electron-photon work-energy relationship (equation) depends on the periodic varying features of the velocity and momentum of the electrons that emit photons (momentum is a quantity of motion of a moving object, measured as its mass multiplied by its velocity). In fact, these features are the prerequisite or indispensable condition for finding and establishing this relationship (that is to say, without revealing these features, it is definitely impossible to find and establish this relationship). Then what are these features? How to reveal them? And why are they wonderfully attractive?

Specifically, these features are entirely determined by the new law (that is, the law of an orbiting electron with periodic impulses emitting photons, as seen above), through the following three crucial aspects. The first aspect:

this law reveals that, when an electron emitting photons is in the state of steady motion (this occurs when an external electric potential difference is maintained at a certain fixed value, such as 12 volts on our vehicle lights and 110 or 220 volts on our family light bulbs), its velocity and momentum return to their same relationships with respect to the position of the electron's equilibrium orbit after one complete impulse period (the velocity and momentum of the electron at points A and E in Fig. 6.3, for example). This revealing shows that at the starting and ending points (points A and E in Fig. 6.3, for instance) over one complete impulse period of the electron, its two velocities are not only equal to each other in value, but also equal to the value of its orbiting velocity (which is the velocity along the direction of the position of the equilibrium orbit of the electron). And these relationships also exist in the momentum of the electron. The second aspect: the essence of the *periodic* impulses of the electron determines that its average velocity over one complete impulse period is equal to its orbiting velocity in value, so is the momentum of the electron. The above two aspects together determine the third aspect (based on the basic principle of vector adding up or superposition; vector is a quantity that has both magnitude (size) and direction, such as force, velocity and momentum): over one complete impulse period of the electron, its velocity *change* is equal to its average velocity in value, so is the momentum of the electron. What should be pointed out is that over the range of one complete impulse period of the electron, the *change* in its velocity and momentum is due to the *change* in direction, rather than in magnitude or value.

THE CERTAINTY MECHANISM BEHIND THE UNCERTAINTY PRINCIPLE

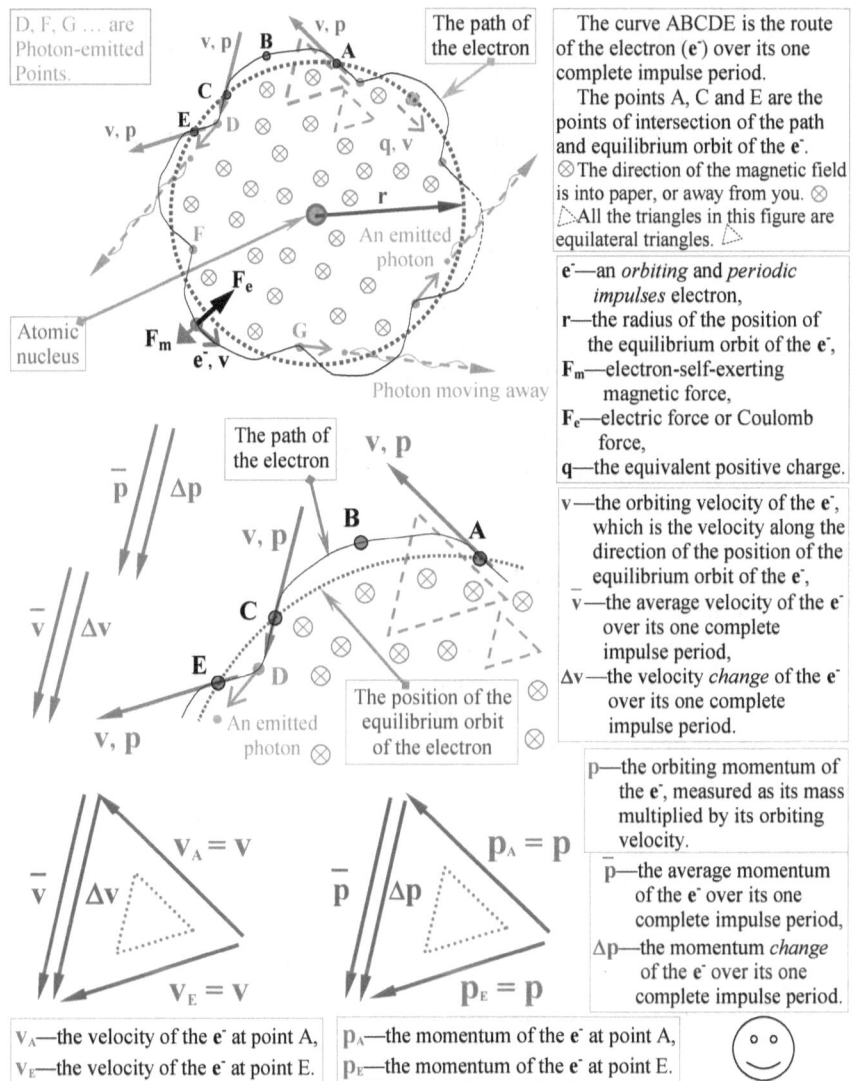

Figure 6.3, the periodic varying features of the velocity and momentum of the electrons that emit photons (using one electron as an example in this illustration).

Collectively, these three aspects are the periodic varying features of the velocity and momentum of the electrons that emit photons: over one complete impulse period of such an electron, its average velocity, velocity *change* and orbiting velocity are equal to each other in value, so is the relationship in the momentum of the electron. As a result, (over the range of one complete impulse period of the electron) its two velocities at starting and ending points, with its average velocity or velocity *change*, constitute an equilateral triangle—a triangle in which all three sides are equal (Fig. 6.3); so is the momentum of the electron. What should be aware is that these periodic varying features are the most important attributes of the electrons emitting photons in the state of steady motion; this state exists when an external electric potential difference is kept at a stable or fixed value (such examples include 12 volts on our vehicle lights and 110 or 220 volts on our family light bulbs)—that is, this state represents the normal state or typical situation experienced in our daily life and in our practical applications. And what should be noticed is that these periodic varying features are the necessary condition or indispensable requirement for the establishment of electron-photon work-energy relationship (equation).

Commentator: up to here, particularly on seeing the two equilateral triangles above (Fig. 6.3), which illustrate the periodic varying features of the velocity and momentum of the electrons that emit photons, some readers may be amazed: how simple these features are! How wonderful these features are! Yes, they are. In fact, what makes science really wonderful lies with that the fundamental truths revealed by it are amazingly simple; there are quite a lot of such instances in science. Newton's second law is simple, but many people admire and appreciate its wonderful implications (this law is the famous $F = m\mathbf{a}$, where m is the mass of an object, $\mathbf{a}$ is the acceleration of the object, and F is the net force acting on the object). Einstein's famous mass-energy equivalence equation is simple, but many people admire and appreciate its wonderful implications (this equation is $E = mc^2$, where c is the speed of light, m is the rest mass of an object, and E is the rest energy of the object). The famous Planck-Einstein equation is simple, but many people admire and appreciate its wonderful implications (this equation is $E = hf$, where h is Planck constant, f is the frequency of a photon, and E is the energy of the photon). The structure of DNA is simple, but many people admire and appreciate its wonderful implications ... and so on. In science, one of the most noticeable marks is: wonderfulness comes from simplicity; simplicity originates from the truth revealed by science—"truth is simple" has become one of the well-known principles. Moreover, the typical or essential nature of truth is simple, which has been clearly substantiated or impartially witnessed by a lot of such instances in science, like those just mentioned.

With these famous facts as an inspiration or encouragement, as to the periodic varying features of the velocity and momentum of the electrons that emit photons, the simplicity of these features seems to be, at least in an intuitive sense, a possible indication of their wonderful implications. Of course, whether these features are really wonderful or not, it must be eventually determined by such a crucial and decisive criterion: whether the application of these features can lead to the conclusions that agree with experimental results or known facts, which will be mentioned as an extremely important issue after a short while.

Having been familiar with the periodic varying features of the velocity and momentum of the electrons that emit photons, we get prepared to enter and experience electron-photon work-energy relationship (which is also called electron-photon work-energy equation, because this relationship is expressed with this equation). (I do hope dear readers will find that this relationship is interesting and attractive, thus enjoyable!)

Briefly, this relationship is the result of an application of the new law (that is, the law of an orbiting electron with periodic impulses emitting photons; OEPIEP law, for short). Specifically, this relationship is collectively determined by three factors. First factor: the periodic varying features of the velocity and momentum of the electrons that emit photons, as just introduced above. Since these features are totally determined by OEPIEP law, this factor is a product of OEPIEP law. The second factor: one photon is emitted from the electron over its one complete impulse period, as revealed by OEPIEP law; so this factor is also a direct component of OEPIEP law. These two factors can therefore be represented with OEPIEP law. The third factor: the famous Planck-Einstein equation $E = hf$ (where h is Planck constant, f is the frequency of a photon, and E is the energy of the photon). Thus, one can simply say that electron-photon work-energy relationship is ultimately the combination of OEPIEP law and Planck-Einstein equation; or more concisely, this relationship is the result of an application of OEPIEP law, to be even simpler. In addition, what should be pointed out is: this relationship obeys energy conservation law (one of the most fundamental principles in science, being familiar even to the students of middle or high schools. About this law, please see the Glossary of this book if necessary), which is reflected in the fact that at the starting and ending points (points A and E, for example; Fig. 6.2, P. 173 or Fig. 6.3, P. 183) over one complete impulse period of the electron, its total energy is the same.

With these three factors, electron-photon work-energy relationship (equation) was established recently (just a few years ago by me, the author of this book). This relationship shows that: the amount of <u>work</u> accepted by an electron over its one impulse period is equal to the <u>energy</u> of the photon emitted by the electron in the impulse period. Therefore, electron-photon work-energy relationship reveals the relationship between the <u>work</u> accepted by an electron and the <u>energy</u> of its emitting photons, thus represents the most essential connection between an electron and its emitting photons. As a result—as a clear and definite result, also as a crucially important result, this relationship turns out to be the indispensable and unavoidable bridge between an electron and its emitting photons. (Related question and answer: what is the key to grasping electron-photon work-energy relationship easily and quickly? Answer: please notice that the work accepted by an electron becomes the energy of the photons emitted by the electron.) (*By the way, the establishment of electron-photon work-energy relationship has also naturally brought out the sincere voice from the heart of the author of this book: the famous Planck-Einstein equation should be given considerable credit for the birth of this relationship, because this famous equation has played an indispensable role to the establishment of this relationship, as just mentioned above. That is to say, though electron-photon work-energy relationship has been established by me, the great contributions from the founders of the famous Planck-Einstein equation, Max Planck and Albert Einstein, not only shouldn't be forgotten, but also should be remembered forever! Therefore, the newly established electron-photon work-energy relationship, on behalf of me, wants to express its and my genuine thanks to Planck and Einstein, the two great scientists of the 20th century.)

When people see a new theory, their first reaction, or the first question coming into their minds, is usually as: has this new theory been verified? Or does this new theory agree with experimental results? (Undoubtedly, this kind of question is not only crucially important, but also definitely necessary in science; it is also a clear reflection of one's rational, correct and wise mode of thinking.) Certainly, this kind of question is equally applicable to the newly established electron-photon work-energy relationship. That is, this relationship has to face the crucially important question like: has electron-photon work-energy relationship been verified or confirmed? Or does this relationship agree with experimental results? The answer is: YES. In fact, this YES is strongly supported and clearly shown by the hard evidence coming from more than one angle, which is reflected in the pieces of concrete information in the following several paragraphs.

The newly established electron-photon work-energy relationship has been extensively confirmed and intensively verified from different angles. First angle: this relationship has been confirmed through deriving out Joule's law from it—that is, the result obtained from this relationship is exactly identical with Joule's law about power output, which is that power output is proportional to the square of the value of an external electric potential difference, and inversely proportional to electric resistance. This accurate agreement has profound implications with great significance! It shows that the newly established electron-photon work-energy relationship is totally consistent with the famous, fully accepted, fundamentally important Joule's law (which is a highly important physical law in the textbooks written for middle or high schools. That is to say, this law is pretty familiar to today's educated people). It provides an effective way for one to grasp electron-photon work-energy relationship tangibly and firmly, because Joule's law is so realistic and touchable in our daily life. It eventually realizes the essential connection between the micro- and macro-worlds by identifying that this newly found relationship turns out to be the mechanism underlying Joule's law. It is also an explicit demonstration that this relationship is exactly the mechanism of how input electrical energy is converted into released photon energy.

Not only that, what directly comes along with this accurate agreement is the specific confirmation of the two crucial components of electron-photon work-energy relationship. One is the confirmation of the newly discovered law of an orbiting electron with periodic impulses emitting photons (simply because electron-photon work-energy relationship is a direct application of this law, as clearly pointed out above); this confirmation shows that the result obtained from this law agrees with the universally acknowledged fact—Joule's law very well. Thus this confirmation can enable one to be confident in the validity and reliability of this newly discovered fundamental physical law from the angle of the famous, fully accepted, fundamentally important Joule's law. Another is the confirmation to the above-mentioned periodic varying features of the velocity and momentum of the electrons that emit photons (because electron-photon work-energy relationship is the direct result of a basic application of these features, as just mentioned above). This confirmation tells us that the result obtained by applying these features is accurately consistent with the well-known fact—Joule's law. With this confirmation as hard evidence, we can therefore be quite confident that we have gotten the right picture from the description of these features. With the strong support of this hard evidence, these features appear even more wonderful and attractive (than they were first presented above)!

Second angle: electron-photon work-energy relationship has been further tested through the verification of its directly causing electron-photon wavelength relationship or equation. (Note: this wavelength relationship comes from that the frequency of a photon is equal to the impulse frequency of the electron that emits *the* photon, as clearly shown in electron-photon work-energy relationship. Specifically, this wavelength relationship, established just a few years ago by the author of this book, reveals and determines, *for the first time in the history of science*, the connection in wavelength between an electron and its emitting photons as follows. The ratio of the wavelength of an electron to the wavelength of its emitting photons is equal to the ratio of the orbiting velocity of the electron to the speed of light.) The result in this verification shows: the wavelength of the electrons calculated with electron-photon wavelength relationship perfectly agrees with the value measured in the famous, fully recognized, Davisson-Germer experiment! (For this experiment, Clinton Davisson was awarded the Nobel Prize for physics.) (Again, this verification is also the verification of the two crucial components of electron-photon work-energy relationship: one is the newly discovered law of an orbiting electron with periodic impulses emitting photons; another is the periodic varying features of the velocity and momentum of the electrons that emit photons. Thus this verification is a further demonstration that the result obtained by applying these two crucial components agrees with the experimental result very well, which can enable one to be more confident in the validity and reliability of these two crucial components.) All in all, with the above confirmations and verifications as experimental evidence or as hard evidence, the validity and reliability of electron-photon work-energy relationship are quite positive! So it is rather rational that we ought to be fully, at least highly, confident in this relationship, as well as its profound and innovative implications.

What follows electron-photon work-energy relationship are its profound and wonderful, also revolutionary, implications! First, this relationship turns out to be the essential link or connection between an electron and its emitting photons. In the past, it was known that photons are emitted from electrons (especially through the germination and development of quantum physics in the early 20th century, as clearly mentioned earlier in this chapter), but the essential connection between an electron and its emitting photons was not discovered. Now, with the establishment of electron-photon work-energy relationship, it is clear (it is fundamentally thus essentially clear) that this essential connection turns out to be: the amount of <u>work</u> accepted by an electron over its one impulse period is equal to the <u>energy</u> of the photon emitted by the electron in the impulse period. Therefore, electron-photon work-energy relationship reveals that an electron and its emitting photons are inherently connected by the work-energy relationship that exists between them; that is, this relationship is the indispensable bridge between an electron and its emitting photons. (Commentator: quite noticeably, this implication, because it is about the revealed fact that electron-photon work-energy relationship turns out to be the essential connection between an electron and its emitting photons, can easily make or help one get a very enlightening inspiration with great importance and significance. This inspiration behaves as: it is of crucial importance finding out the essential connection between different, but intrinsically related, phenomena in science, because knowing such an essential connection is one of the most important goals of science. Moreover, because both electrons and photons are the fundamentally important phenomena or subjects in science and in the process of science development and advancement, knowing the essential connection between electrons and their emitting photons is of fundamental importance and great significance. With such a fundamentally important and meaningful goal as an explicit reminder, it becomes clear that the birth of electron-photon work-energy relationship is indeed a historic milestone in the development of physics. As a result—as a magnificent and wonderful result, the establishment of electron-photon work-energy relationship, when viewed from the perspective of science development or progress, is really a great and historic event.)

In addition, when viewed from a larger perspective or wider angle, this implication, because it is about the truth that work-energy relationship is still the essential connection between an electron and its emitting photons, has injected remarkably new significance into the well-known, traditional or "old" principle of work-energy relationship. With and for this implication, the "old" principle of work-energy relationship is markedly and wonderfully rejuvenated from now on. (Reviewer: 'work-energy relationship', what a fundamental idea in science! What an important and familiar concept in physics! What a crucial relationship in physics!) As a result, this marked and wonderful rejuvenation can and will make the greatly important principle of work-energy relationship become even more important.

Second, electron-photon work-energy relationship essentially and radically extends the famous Planck-Einstein equation (again, this equation is $E = hf$, where h is Planck constant, f is the frequency of a photon, and E is the energy of the photon). Quite explicitly and obviously, this famous equation is only connected with a photon, but has nothing to do with the electron that emits *the* photon; in fact, this equation is utterly unable to connect a photon with the electron that emits *the* photon at all. The newly established electron-photon work-energy relationship accomplishes the essential and radical extension of this famous equation, by connecting the parameters of a photon, such as its frequency, with those of the electron emitting *the* photon, such as the impulse period and orbiting velocity of the electron. In other words, electron-photon work-energy relationship revolutionizes the famous Planck-Einstein equation by connecting the parameters of a photon with the parameters of the electron emitting the photon. With this essential extension and connection, photons are no longer like a tree without roots, or water without a source. (Related question and answer: what is the biggest or most noticeable difference between electron-photon work-energy relationship and the famous Planck-Einstein equation? Answer: the former includes both the electron and its emitting photons, whereas the latter is merely able to involve photons.) (Commentator: therefore, from the large perspective of the fundamental nature of science, it seems rational and objective to see and say that the birth of electron-photon work-energy relationship is a great step forward from photons to the electron that emits them in the advancement of physics. This great step is a giant step from phenomenon to mechanism, a giant step from effect to cause, a giant step from appearance to essence! Moreover, this great step is also the necessary giant step for the further developments of physics.)

Third implication: electron-photon work-energy relationship finally unveils the mysterious veil of *why* the energy of photons is proportional to their frequency. (Though the famous Planck-Einstein equation, again which is E = hf —where h is Planck constant, and E is the energy of a photon with frequency f, tells us that the energy of photons is proportional to their frequency, it doesn't and can't tell us *why;* in fact, this equation doesn't and can't know this *why*.) Electron-photon work-energy relationship reveals that the frequency of a photon is equal to the impulse frequency of the electron emitting *the* photon. This means that a photon with a higher frequency is emitted from an electron with a shorter impulse period, because the impulse frequency of the electron is inversely proportional to its impulse period (that is, its impulse frequency times its impulse period equals one). As shown in the newly established and verified electron-photon work-energy relationship, the shorter the impulse period of the electron, the more work it accepts over its one impulse period; thus the more the energy of the photon emitted by the electron. On the other hand, a photon with a lower frequency is emitted from an electron with a longer impulse period; the longer the impulse period of the electron, the less work it accepts over its one impulse period; thus the less the energy of the photon emitted by the electron (also as revealed or shown in electron-photon work-energy relationship). What you've read above are the secrets of *why* the energy of a photon is proportional to its frequency, or the secrets of *why* the energy of photons is proportional to their frequency! Therefore, electron-photon work-energy relationship, and only this relationship, ultimately solves the long-standing, fundamentally important puzzle of *why* the energy of photons is proportional to their frequency. (Narrator: the mysterious veil on the famous Planck-Einstein equation has been finally unveiled at the age of more than one century old. After the unveiling, this famous equation appears even more beautiful and charming, because this unveiling is the uncovering the answer to the long-standing big puzzle of *why* the energy of photons is proportional to their frequency; because this unveiling is the revealing of the most essential connection or link between an electron and its emitting photons, the work-energy relationship that exists between an electron and its emitting photons.)

(Along with this wonderful implication, it seems better that I should provide a relevant clarification here for avoiding possible confusion. This clarification is: in the famous model of the hydrogen atom built by Niels Bohr in 1913, what was attempted to relate the frequency of photons was

THE CERTAINTY MECHANISM BEHIND THE UNCERTAINTY PRINCIPLE

the *orbital* frequency, rather than the *impulse* frequency, of the electron described in that model; not to mention that the concept or idea about the *impulse* frequency of the electron emitting photons never appeared in that model! Moreover, if one thinks deeply, it seems not difficult that he or she can clearly perceive or realize: the electron that appears in that model is actually unable to accept the work done on it over an entire *orbital* period—no wonder that model doesn't and can't reflect the work-energy relationship between an electron and its emitting photons. Not only that, if one keeps thinking over and asking the fundamentally important question "Since that model admits photons are emitted from an electron, how do photons get their velocity c—the speed of light, from the electron emitting them?" the fatal flaw of that model will appear immediately, because that model doesn't and can't have the function or ability to answer this question at all. In fact, that model utterly skips over this question, being a noticeable or clear *fact* to the people who have been familiar with that model.)

Fourth, last but not least—in fact, *this last* is perhaps the most spectacular and brilliant highlight of the four profound and wonderful implications discussed up to here, because *this last* is the central task of this chapter, also the destination of this journey. THIS LAST is, is, is … is that the newly established electron-photon work-energy relationship turns out to be the *certainty* mechanism behind the *uncertainty* principle. Why? How? The answers to them are the very subject to be coming up. The leading role of this chapter is coming up! This leading role is about to be unfolding its splendid and brilliant fascinations! We are going to enter the climax or focus of this journey.

After all the necessary preparations above, we are well equipped to go to the central task of this chapter, which, and again, is a brief introduction of revealing and determining the *certainty* mechanism behind the famous *uncertainty* principle. Thus, the fascinating certainty mechanism behind the wonderful and amazing uncertainty principle is about to spread out its great charm to all of us! Dear readers, are you ready? Let us go!

Since the uncertainty principle is about the electrons that emit photons (this is determined by the origin of this principle, as analyzed and pointed out earlier), and since electron-photon work-energy relationship is the most essential connection between an electron and its emitting photons, it seems reasonable, at least not irrational, if one thinks of or asks the question like: does this relationship have any link or connection with the *certainty* mechanism behind the *uncertainty* principle? Or more specifically thus more directly, is this relationship the *certainty* mechanism behind the *uncertainty* principle? The answer to this question is positive and definite: the newly established and verified electron-photon work-energy relationship turns out to be indeed the *certainty* mechanism behind the *uncertainty* principle (that is, this relationship turns out to be exactly the same thing as this *certainty* mechanism)! Of course, such a short answer cannot allow dear readers to view and enjoy the captivating charms of this certainty mechanism. So it seems better to go a little further along the route leading to this answer; this route will guide us to experience and appreciate the wonderful fascinations of this certainty mechanism!

The task of revealing and determining the certainty mechanism behind the uncertainty in the momentum and position of the electrons that emit photons was finished (by me, the author of this book, soon after I had established electron-photon work-energy relationship) through the following route (momentum is a quantity of motion of a moving object, measured as its mass multiplied by its velocity). First, it was the combination of the two factors (one is electron-photon work-energy relationship, another is the periodic varying features of the velocity and momentum of the electrons emitting photons) that determined the third factor of an electron that emits photons: the value or size of the *change* in the momentum and wavelength of the electron over its one complete impulse period. Such a determination then established the certainty mechanism equation of the uncertainty in the momentum and position of the electron. (The most important feature of this equation is: over the range of one complete impulse period of the electron, the three products of its two crucial quantities, momentum and wavelength, along with their *change*—its momentum *change* times its wavelength *change*, its momentum *change* times its wavelength, and its momentum times its wavelength, are not only equal to each other in value, but also equal to the value of Planck constant h. Note: the product of two quantities, or the product of quantity A and quantity B, or the product of number A and number B, is A times B; for example, the product of 3 and 5 is 15.) Finally, it was this equation that revealed the certainty mechanism behind this uncertainty. (The key to grasping this revealing lies in: the *change* in the momentum and wavelength of the electron corresponds to the *uncertainty* in the momentum and position of the same electron.) Concisely, this equation showed us two clear facts. One is that the certainty mechanism behind this uncertainty is electron-photon work-energy relationship, which is the most essential relationship or link between an electron and its emitting photons. Another is that over the range of one complete impulse period of the electron, the value of this certainty mechanism is equal to that of Planck constant h.

Having known the certainty mechanism behind the uncertainty in the momentum and position of the electrons that emit photons, one may easily think of or naturally ask: what is the certainty mechanism behind the uncertainty in the energy and time of the electrons emitting photons (another common form of the uncertainty principle)? Is there the same certainty mechanism behind the two forms of the uncertainty principle? These questions have been answered by completing the task of revealing and determining the certainty mechanism behind the uncertainty in the energy and time of the electrons that emit photons (this task was also finished by the author of this book, soon after he had established electron-photon work-energy relationship).

This task has been completed through the combination of the two factors too (again, one is electron-photon work-energy relationship; another is the periodic varying features of the velocity and momentum of the electrons emitting photons). This combination then led to the birth of the certainty mechanism equation of the uncertainty in the energy and time of the electrons that emit photons. Finally, it was this equation that revealed the certainty mechanism behind this uncertainty. (The key to grasping this revealing lies with: the *change* in the energy of the electron over the short duration more or less equal to the impulse period of the electron, which is caused by the electron accepting the work done on it, corresponds to the *uncertainty* in the energy and time of the same electron.) Briefly, this equation also showed us two clear facts. One is that the certainty mechanism behind this uncertainty is still electron-photon work-energy relationship, which again is the most essential relationship or link between an electron and its emitting photons. Another is that over the range of one complete impulse period of the electron, the value of this certainty mechanism is equal to that of Planck constant h too. (Commentator: thus, even though the uncertainty principle has two different forms, both have the same certainty mechanism, which can be a further manifestation or clear reflection of the authenticity, validity and reliability of this certainty mechanism. More noticeably, since the uncertainty principle has been prevalently regarded as one of the most fundamental and important theories in physics, undoubtedly, also undeniably, the *certainty* mechanism behind this principle is one of the most fundamental and important mechanisms in physics! More than that, because the uncertainty principle has been widely viewed as one of the most significant and influential concepts in science, clearly, even obviously, also plainly and indisputably, the *certainty* mechanism behind this principle is one of the most significant and influential mechanisms in science!)

Eventually, when the famous uncertainty principle was at the age of nearly ninety years old, its secret veil was finally unveiled—the *certainty* mechanism behind the *uncertainty* principle has been revealed at last. (Commentator: fortunately, this principle was not a bride! Of course, if it had been a bride, probably no bridegroom in the world would have been patient enough to wait for such a long time to unveil her veil after their wedding. But for a highly influential scientific theory like the uncertainty principle, it seems that the later unveiling its secret veil, the more wonderful its marvelous charm is. This seems to be a bit like wine: a bottle of old wine is more tasteful and mellower than a new one.) In this sense, the uncertainty principle should be happy for itself. In this sense, the uncertainty principle ought to congratulate on itself. In this sense, the uncertainty principle must celebrate itself! Moreover, and in a broad sense, it seems acceptable if the uncertainty principle wants to invite all the people in the world, especially the respected and related experts in physics, to have a grand and solemn celebration of this great and historic revealing! (Most probably, the uncertainty principle will provide delicious food and excellent wine for all of us in such a spectacular, splendid, and wonderful occasion.)

At this moment, at the beautiful and marvelous moment when the *certainty* mechanism behind the *uncertainty* principle has just unfolded before us, some readers may be amazed at the wonderful charms and fascinations of this *certainty* mechanism. What does this amazement mean? It can be, at least in a sense, an indication that there seems to be a need of some extra information that can specifically and explicitly help them sense and understand this certainty mechanism well, so that they can have a clearer and better view of its wonderful charms and fascinations. How to accomplish that then? We shall let this newly revealed certainty mechanism become much more noticeable by making it manifest more clearly and explicitly. Such a task is about to be carried out through providing some extra clues or evidence that can substantially help dear readers have a tangible and explicit understanding of this certainty mechanism. These clues or evidence can be as wonderfully beautiful as this certainty mechanism itself. So let us go now; let us go towards the fascinating and wonderful beauty of these clues or evidence.

But immediately before starting out, please allow me to remind two things. First, the *certainty* mechanism behind the *uncertainty* principle is electron-photon work-energy relationship; that is to say, once one has grasped this relationship from the standpoint of the uncertainty principle, he or she will actually grasp this certainty mechanism. Second, over the range of one complete impulse period of the electron, the value of this certainty mechanism is equal to that of Planck constant h, which has just been presented above. Correspondingly, this journey has two clear and definite directions. One is towards: what is the key to effectively comprehending electron-photon work-energy relationship from the perspective of the uncertainty principle. Another is towards looking for the answer to such an obviously or remarkably important question: does electron-photon work-energy relationship (that is, the *certainty* mechanism behind the *uncertainty* principle) turn out to be also the mechanism and essence of Planck constant, considering the universally recognized fact that statistically, the average value of the uncertainties of the uncertainty principle is more or less equal to the value of Planck constant h? (If the answer to this question is YES, then we will find out a simple and easy method of understanding this certainty mechanism tangibly, clearly and quickly, because Planck constant h appears in either of the two mathematical expressions of the uncertainties that are defined in the uncertainty principle.)

Let us start towards the first direction: how to grasp electron-photon work-energy relationship effectively from the perspective of the uncertainty principle. Along this direction, there are three viewing angles. The first viewing angle is from: why the two variables (momentum and position, for example) in the uncertainties of the uncertainty principle have to be joined in pair by multiplication symbol. The uncertainty principle has two forms of uncertainty. One is the uncertainty in momentum and position; another is the uncertainty in energy and time. These two forms of uncertainty have such a common feature: momentum and position are coupled with *multiplication symbol;* energy and time are connected with *multiplication symbol* too. That is, all the variables in the uncertainties of the uncertainty principle appear not only pair by pair, but also in the manner of the product of two variables—the product of momentum and position; the product of energy and time (note: the product of two variables, or the product of variable A and variable B, or the product of number A and number B, is A times B; for example, the product of 3 and 7 is 21). If one thinks deeply, it seems quite reasonable that he or she can naturally or easily ask: why? Is there a deeper reason why these variables have to be jointed in such a way (instead of the two variables appearing in other manners or forms)? Or what is the underlying reason behind this common feature?

After revealing the *certainty* mechanism behind the *uncertainty* principle, the answer to these questions has become clear: the underlying reason behind this common feature turns out to be electron-photon work-energy relationship, the *certainty* mechanism behind the *uncertainty* principle. For a better understanding of this answer, two important aspects are worthy to be pointed out thus to be noticed. First aspect, only these variables have this common feature, can they finish their common "task" together: causing the electron that emits photons to be able to accept the amount of work (over the duration of one complete impulse period of the same electron) equal to the energy of its emitting photon. That is, the completion of such a common "task" simultaneously requires the joint function of both variables—momentum and position, or energy and time. This is a bit like that the growth of a crop requires the joint function of water, temperature and nutrient at the same time. Second aspect, only with this common feature, can the two variables work together and finish the "job" that cannot be done by either of these two variables alone. One can easily perceive and grasp this aspect through the concrete working function of tongs or Chinese chopsticks as a tangible and vivid analogy. And so, this common feature turns out to be just the indispensable requirement for materializing the *certainty* mechanism behind the *uncertainty* principle. In other words, this common feature turns out to be a clear clue or noticeable signal that may, at least in a sense, indicate the existence of electron-photon work-energy relationship. Such an indication can substantially help one comprehend this relationship—the *certainty* mechanism behind the *uncertainty* principle.

The second viewing angle was from the collective reflection of the opinions on the uncertainty principle from some great scientists in the 20th century, including Heisenberg himself—the developer of this famous principle, Einstein, Planck, and Schrodinger, etc. Let us together review these opinions. First, let us go to the opinion of Heisenberg. In the process of developing the uncertainty principle, he clearly realized the limits of physical knowledge, although this clear realization was a great shock to him (according to his interpretation, the uncertainties in the uncertainty principle were inherent in quantum mechanics, instead of the fault of the experimenter). Second: the opinion from Einstein. He never believed that the uncertainty principle was a complete theory. He insisted that there must be some underlying reason that produces the statistical behavior displayed in the uncertainty principle; and he had faith that some future theory could reveal this reason. Third: the opinion of Planck and Schrodinger. They both had a strong objection to the indeterministic, statistical description of the uncertainty principle; they believed that this sort of description was due to a yet undiscovered law. Fourth: the historical fact, which was the collective reaction of many scientists at the time when this principle was newly born. When the uncertainty principle was initially proposed, many experts in physics resisted it based on that this principle did not have a mechanism—that is, these experts believed or expected that there should be a mechanism behind this principle.

Now, after the revealing of the *certainty* mechanism behind the *uncertainty* principle, the clearly shown or explicit existence of this certainty mechanism turns out to be, at least in concept or in principle, actually consistent with all the opinions mentioned above. For example, *after revealing this certainty mechanism*, to Heisenberg, his realization of the limitations of physical knowledge turned out to be correct, at least in a large sense (because this certainty mechanism had not been revealed in the past). The inability to reveal this certainty mechanism was, of course, a specific manifestation of the limitations of physical knowledge. *After revealing this certainty mechanism*, to Einstein, his belief that the uncertainty principle was not a complete theory turned out to be correct, at least by and large, in terms of this certainty mechanism had not been revealed in his times. And the underlying reason expected by him turned out to be just this certainty mechanism. *After revealing this certainty mechanism*, to Planck and Schrodinger, a yet undiscovered law turned out to be exactly this certainty mechanism. All in all, this kind of widespread and profound consistency can serve as a clear clue that may indicate, from the angle of the uncertainty principle, the existence of this certainty mechanism. Such an indication can considerably help one grasp this certainty mechanism, being similar to the indication from the first viewing angle—as just discussed above.

The third viewing angle is from the prominent, well-known and fully recognized feature of the uncertainties themselves of the uncertainty principle. This feature is that statistically, the average value of these uncertainties is always quite comparable to the value of a certain quantity, which is known to be Planck constant h. Such a feature can explicitly help one perceive the solid existence of the *certainty* mechanism behind the *uncertainty* principle, with the simple reasoning like: IF this *certainty* mechanism did not exist, then the average value of these uncertainties would be simply equal to zero! In other words, this prominent feature—also a fully accepted *fact*, which is that the average value of these uncertainties is equal to a certain positive number (particularly this positive number is practically equal to the value of the extremely important Planck constant, a crucially important constant in quantum physics, also a fundamentally important constant in physics and in science), turns out to be a good indication of the noticeable existence of this certainty mechanism. Such an indication can tangibly help one realize a manifest, eye-catching clue that points to the explicit existence of this certainty mechanism. Such a realization, needless to say, can definitely help one ponder and understand this *certainty* mechanism, at least in some sense or from a certain angle. (Commentator: besides, this realization can considerably reduce the psychological surprise or shock of some people, especially some of those respected professional people in physics—they are very familiar with this well-known and fully recognized feature, towards the revealing of the *certainty* mechanism behind the *uncertainty* principle. Clearly, even obviously, this effect is quite helpful for them to face this *certainty* mechanism directly and rationally, thus grasp it easily and quickly.)

All right, along the journey in the first direction (which is: what is the key to effectively comprehending electron-photon work-energy relationship from the perspective of the uncertainty principle), and through the three viewing angles above, one can clearly see thus definitely realize the solid existence of electron-photon work-energy relationship. Quite obviously, this realization can enable one to comprehend the *certainty* mechanism behind the *uncertainty* principle well, because this *certainty* mechanism is also electron-photon work-energy relationship, as presented above. This comprehension, in turn, can make some, even many, attentive and sharp readers explicitly and surely perceive, though perhaps somewhat unexpectedly or amazingly: it turns out that they are actually very close to the great charms and brilliant fascinations of this *certainty* mechanism whenever the famous and wonderful *uncertainty* principle appears before them. (Commentator: with this satisfying perception, it seems quite reasonable or pretty realistic if one comes to the rational conclusion that some, even most, readers might have been well aware that the journey along this direction was really interesting and attractive.)

Yes! The beautiful and fine sights along the first direction are certainly attractive and arresting, thus could draw people lingering on them and hard to forget. But the wonderful and superb views in the second direction will surely be fascinating and captivating, thereby to be inviting people to yearn for excitedly. After finishing the journey in the first direction, let us turn to the path along the second direction. In fact, what dear readers will experience in the coming direction can be even more interesting and attractive (than in the first direction). This is because the second direction will guide dear readers onto the road of seeing the clear answer to the crucially important question: does electron-photon work-energy relationship (which is also the *certainty* mechanism behind the *uncertainty* principle) turn out to be the mechanism and essence of Planck constant too?

The answer to this question is positive and definite; that is, electron-photon work-energy relationship turns out to be also the mechanism and essence of Planck constant! This answer is obviously a considerable help for one to have a good grasp of the *certainty* mechanism behind the *uncertainty* principle, because of the following two clear reasons. One is that statistically, the average value of the uncertainties of the uncertainty principle is more or less equal to the value of Planck constant h—being one of the most prominent features of these uncertainties, also being a fully recognized *fact*, as just mentioned above (along with the eye-catching reminder that Planck constant h appears in either of the two mathematical expressions of the uncertainties of the uncertainty principle). Another is that electron-photon work-energy relationship has been revealed or shown to be this *certainty* mechanism.

These two clear reasons determine and demonstrate that the above answer is not only certainly important, but also surely wonderful and attractive, because it provides a simple and effective way that can explicitly help one identify and grasp the *certainty* mechanism behind the *uncertainty* principle clearly and easily. However, the specific process leading to the answer above, as readers will clearly see, will be much more excitingly wonderful and attractive (than this answer itself). Thus, such a specific process will be our attention or itinerary in the wonderful and delightful tour along this direction. Let us start on this wonderful and delightful tour now!

Let us start from the great significance of the highly important Planck constant. This constant (being represented with h in this book) is the most pivotal constant in quantum physics, also one of the three fundamental constants in physics (the other two are the universal gravitational constant G and the speed of light c). Broadly speaking, Planck constant h is also a fundamentally important constant in science.

After its birth in 1900 (along with Planck's pioneering quantum theory or quantum hypothesis for interpreting the energy radiated from a hot object or body, like an oven or a star), Planck constant h had soon become the core constant in quantum physics during its golden developing period from 1900 to the 1930s. This was demonstrated in a host of facts that Planck constant, being a hub constant, was *indispensable* to almost all the central theories of quantum physics. It was *indispensable* to Einstein's interpretation of the photoelectric effect in 1905. It was *indispensable* to the famous model of the hydrogen atom (put forward by Niels Bohr in 1913). It was *indispensable* to the wavelength equation for electrons (suggested by Louis de Broglie in 1924). It was *indispensable* to quantum mechanics (developed by Schrodinger and Heisenberg in 1925). It was *indispensable* to the famous Davisson-Germer experiment for establishing the wave nature of the electrons (in 1927). It was *indispensable* to the uncertainty principle (established by Heisenberg in 1927) ... etc. All in all, Planck constant h, quite similar to a strong and crucially necessary glue, appears in all the important equations of quantum physics; these equations would utterly collapse if this constant were taken away from them. Therefore, Planck constant h, being a crucially indispensable and extraordinarily important core constant, plays a central and hub role in quantum physics. Accordingly, the great importance of this constant cannot be overestimated in the realm of quantum physics!

However (this is a big and heavy HOWEVER with profound implications and great significance!), there are several crucially important questions that have been hovering over Planck constant since its birth. These long-term unanswered big questions behave as: since Planck constant is inseparable from photons (for example, in the famous Planck-Einstein equation mentioned several times above, Planck constant h is the coefficient before the frequency of a photon), and since photons are emitted from electrons, does Planck constant have the essential implication to the electrons that emit photons? What crucially bridging role does Planck constant play in the relationship between an electron and its emitting photons?

What is the underlying reason that causes Planck constant to be a constant (an invariable quantity)? Or what is the inherent mechanism or essence that tells us *why* Planck constant has the same value in different situations? These long-standing crucially or fundamentally important questions have been collectively and persistently sending us such a clear signal: the mechanism thus essence behind Planck constant has not been revealed yet, which indicates that fundamentally speaking, this constant is still a great and deep mystery to all of us! This clear signal thus reminds us: the really great and wonderful charm behind Planck constant has not been dug out and noticed yet; the truly splendid and marvelous charm of Planck constant is still unknown to our human beings.

We should and must dig out the brilliant and splendid charm of the highly important Planck constant with profound implications and great significance! The way to do this, and the only way to do this, is to reveal the mechanism and essence behind Planck constant. (Related question and answer: what is the noticeable and crucial difference *before* and *after* revealing the mechanism and essence behind Planck constant? Answer: before this revealing, Planck constant was only related to photons, but had nothing to do with the electrons that emit photons. After this revealing, as dear readers will see soon, Planck constant is not only related to photons but also linked to the electrons that emit photons; moreover, Planck constant, after this revealing, turns out to be the indispensable bridge between an electron and its emitting photons.)

Briefly, the task of revealing the mechanism and essence behind Planck constant was finished just a few years ago (by me, the author of this book, soon after I had established electron-photon work-energy relationship). Overall, this task was carried out along such a route: electron-photon work-energy relationship (equation), which is the most essential relationship or connection between an electron and its emitting photons, as has been discussed above → Planck constant equation → the mechanism and essence behind Planck constant.

Specifically, Planck constant equation was obtained from electron-photon work-energy relationship (equation) through a simple mathematical transformation. Because the frequency of a photon is equal to the impulse frequency of the electron emitting *the* photon (as pointed out earlier), whereas the impulse frequency of the electron is the inverse of its impulse period (that is, the impulse frequency times the impulse period equals 1; the inverse of 5 is 1/5, for example), the frequency of a photon in electron-photon work-energy relationship was therefore replaced with the inverse of the impulse period of the electron emitting *the* photon. After this replacement, then through a simple rearrangement so that Planck constant h was put alone in the equation, this equation thus became Planck constant

THE CERTAINTY MECHANISM BEHIND THE UNCERTAINTY PRINCIPLE

equation (the meaning of Planck constant equation will be presented in the coming paragraph as the mechanism and essence of Planck constant). What should be emphasized or pointed out is: since Planck constant equation came from electron-photon work-energy relationship, this essentially important relationship is forever kept and maintained in Planck constant equation, because the above replacement did not result in the loss of this relationship at all. In addition, what should be noticed is that, except the replacement above, all the other aspects of this equation are the same as electron-photon work-energy relationship. (So, here I don't think it's necessary to repeat what has already appeared in this relationship earlier; it's enough only to provide the information related to the fundamental nature of Planck constant.) (Related question and answer: has Planck constant equation been verified or confirmed? Answer: yes; it has been comprehensively verified or confirmed through the verification and confirmation of electron-photon work-energy relationship mentioned earlier in this chapter.)

What Planck constant equation brought to us was the *mechanism* and *essence* of Planck constant h. This equation revealed the mechanism behind Planck constant as that: *over one complete impulse period of an electron that emits photons, the amount of work accepted by the electron times its impulse period equals the value of Planck constant h*. This equation also showed that the essence of Planck constant is electron-photon work-energy relationship: *the <u>work</u> accepted by an electron over its one complete impulse period is equal to the <u>energy</u> of the photon emitted by the electron in this impulse period*. Having seen this mechanism and essence, one can thus explicitly realize such a clear fact: electron-photon work-energy relationship turns out to be also the mechanism and essence of Planck constant indeed.

Then what will come along with this clear fact when viewed from the angle of the uncertainty principle? It is obviously a considerable help for one to chew over and digest the *certainty* mechanism behind the *uncertainty* principle, for two explicit and tangible reasons. One is that this certainty mechanism has been revealed to be electron-photon work-energy relationship. Another is the fully accepted fact that statistically, the average value of the uncertainties of the uncertainty principle is more or less equal to the value of Planck constant, along with the noticeable reminder that either of the two mathematical expressions of these uncertainties contains Planck constant h. (Commentator: having known that the mechanism and essence of Planck constant turns out to be the certainty mechanism behind the uncertainty principle too, different readers might have different reactions. Some readers may realize: since the uncertainty principle has been regarded as a wonderful and fascinating theory, the certainty mechanism behind it must or should be a wonderful and fascinating mechanism;

thus, the mechanism and essence of Planck constant is also wonderful and fascinating. Indeed, such a realization is not only correct, but also makes these readers view and appreciate the wonderful and fascinating charms of this *certainty* mechanism from the angle of the highly important Planck constant.) (Of course, it seems pretty reasonable that no one can totally exclude the possibility that this realization may also lead some readers to perceive: the gentleman who fulfilled the task of revealing the mechanism and essence of Planck constant might be a wonderful and fascinating person. Certainly, you are free to have such a perception; at least I have no objection to it.)

In fact, what the mechanism and essence of Planck constant could bring to us was far more than just an easily perceptible, exceedingly effective way that enables one to see the *certainty* mechanism behind the *uncertainty* principle more clearly and explicitly, thus grasp this certainty mechanism more tangibly and definitely. Other related remarkable consequences, due to having displayed the fundamentally profound and essentially important implications of revealing the mechanism and essence of Planck constant, can be even more wonderful and fascinating than this mechanism and essence! So let us go towards the superb and marvelous charm of these consequences. (Narrator: these remarkable consequences are also an extensive and emphatic manifestation of the *certainty* mechanism behind the *uncertainty* principle, because this certainty mechanism, just like the mechanism and essence of Planck constant, has been proven to be electron-photon work-energy relationship. Such a manifestation is thus a good help for one to view and enjoy this certainty mechanism from a large perspective or a wide angle. So the topic of talking about these consequences doesn't stray from the central subject of this chapter—the *certainty* mechanism behind the *uncertainty* principle.) But due to the limited space of this chapter—because I am concerned that some readers will lose patience if too much space is occupied by dwelling on these consequences, what we can visit will be only the two highlights selected from these remarkable consequences.

First highlight: the mechanism and essence of Planck constant eventually unveils the mystery of *why* Planck constant is the same constant (an invariable quantity) in different situations, by uncovering the characteristic features of the electrons that emit photons as follows. According to electron-photon work-energy relationship (this relationship is the essence of Planck constant, as just presented above), in the situation when the orbiting velocity of an electron is increased, the amount of work accepted by the electron over its one impulse period is thus increased; but the impulse period of the electron is shortened proportionally. As a result, (by Planck constant equation or the mechanism of Planck constant revealed by this equation) the product of this amount of work and this impulse period still remains a constant, and this constant is Planck constant (note: the product of quantity A and quantity B, or the product of number A and number B, is A times B; for example, the product of 4 and 10 is 40). On the other hand, in the situation when the orbiting velocity of an electron is decreased, the amount of work accepted by the electron over its one impulse period is thereby decreased; the impulse period of the electron, however, is proportionally elongated (also according to electron-photon work-energy relationship; again this relationship is the essence of Planck constant). And so, the product of this amount of work and this impulse period still remains a constant too, and this constant is Planck constant (being determined by Planck constant equation or the mechanism of Planck constant revealed by this equation). Of course, it is much easier to realize, from the information above, that in the situation when the orbiting velocity of an electron is unchanged, the product of the amount of work (accepted by the electron over its one impulse period) and the impulse period of the electron stays at a constant, and this constant is Planck constant. Therefore, after revealing the mechanism and essence behind Planck constant, the mystery of *why* Planck constant is a constant is no longer a secret to us, or becomes an open secret to everyone.

The second highlight of these remarkable consequences is: *after* revealing the mechanism and essence of Planck constant h, its function or implication experiences a profound extension or enlargement—the extension from photons to the electrons that emit photons. This extension is an essential and radical change in the role of Planck constant (in contrast with its original or 'traditional' function and meaning, which will be mentioned in the coming paragraph). This crucial extension shows that Planck constant turns out to be the indispensable bridge that connects an electron and its emitting photons; this indispensable bridge is electron-photon work-energy relationship, the most essential link between an electron and its emitting photons. Why is this extension so profound and important? The answer to it is coming to us; and this answer will become contrastingly clear after this extension is pasted onto the background of the 'conventional' interpretation of Planck constant.

The interpretation of Planck constant without this extension, which is also the 'conventional' interpretation of this constant as its original or 'traditional' function and meaning, is incapable of connecting photons with the electrons that emit them. For instance, according to the 'conventional' interpretation in the famous Planck-Einstein equation (which is $E = hf$, where h is Planck constant, f is the frequency of a photon, and E is the energy of the photon), Planck constant h is merely the invariable coefficient in front of the frequency of a photon. That is, Planck constant h is only connected with a photon, but is utterly unable to link up with the electron that emits *the* photon. Quite obviously, this interpretation doesn't and can't reflect electron-photon work-energy relationship: this relationship includes both an electron and its emitting photons, whereas this interpretation merely involves photons. Moreover, any other interpretation of Planck constant, as long as without this extension, is unable to reflect or embody electron-photon work-energy relationship either. What should be noticed is: electron-photon work-energy relationship is the mechanism and essence of Planck constant, also the most essential connection between an electron and its emitting photons, as discussed above. In other words, no interpretation of Planck constant, as long as it doesn't include this extension, can reflect the mechanism and essence of Planck constant! (Thus, without this extension, Planck constant is a bit like a tree without roots, or water without a source.) Therefore, and by contrast, this extension (just mentioned above) is really profound and important, as a remarkable consequence after revealing the mechanism and essence of Planck constant.

(Commentator: this profoundly important extension is also the connection from phenomenon to mechanism, from appearance to essence, and from effect to cause! This connection is of great significance, being clearly and objectively determined by the two basic tasks of science. One is to discover the fundamental connection among the different aspects of the same phenomenon; another is to find out the inherent or essential connection between different, but intrinsically related, phenomena. In this sense, it seems objective and fair to say that this crucial extension is, in fact, a great step forward. In this sense, it seems not inappropriate that we should congratulate on Planck constant—the most centrally important constant in quantum physics, for this crucial extension and for this essential connection.)

Broadly speaking or historically, from the large perspective of the history of science development, what we have just discussed above was also a consolation to the earliest founder of quantum physics, the great scientist Max Planck, because he believed that there could be a better explanation for his formula. (Again, his formula is often referred to as the famous Planck-Einstein equation, which is $E = hf$, where h is Planck constant, f is the frequency of a photon, and E is the energy of the photon.) Yes! Planck constant equation, because it has also revealed the mechanism and essence of his formula—being reflected in the two highlights just mentioned above, is clearly a better explanation for his formula. Yes! Revealing the mechanism and essence of his formula is, of course, a better explanation of his formula. (Commentator: thus, this belief or opinion of Planck turns out to be correct! And more importantly, while his opinion has become history, its far-seeing and profound implications are still of great significance nowadays.)

One of such implications seems to remind us: this opinion of Planck can also serve as a visionary prediction or idea that inspires us to view and appreciate the profound, historic significance of establishing Planck constant equation, as well as revealing the mechanism and essence underlying Planck constant. To dear Planck constant h, it eventually has its own equation at the age of more than one hundred years old. This is a bit like that after having remained single for more than one century, Planck constant ultimately finds out its spouse—Planck constant equation. Congratulations to you, Planck constant! This is also similar to that after having been covered with a mysterious veil for more than one century, this mysterious veil is unveiled at last—the mechanism and essence underlying Planck constant has been revealed with Planck constant equation. So congratulations to you again, Planck constant! Moreover, after this revealing, Planck constant turns out to be much more splendidly charming and fascinating than previously thought. The climax or focus of this splendid charm and fascination lies with: Planck constant turns out to be the indispensable bridge that connects an electron and its emitting photons; this bridge is electron-photon work-energy relationship, which is the most essential link between an electron and its emitting photons. (Narrator: as a result—as a clear and definite result, with the birth of Planck constant equation, and after its revealing the mechanism and essence behind Planck constant, from now on, whenever one thinks of Planck constant, the first thing coming into his or her mind should be electron-photon work-energy relationship. From now on, Planck constant is no longer an ordinary constant; it represents the most essential work-energy relationship between an electron and its emitting photons instead.) (Having said those, I can't help thinking of and admiring the great, far-reaching and keen insight in this opinion of Planck! Having read what I have analyzed and explained above, some, even many, sharp and insightful readers might have clearly realized: this belief or opinion of Planck can't and shouldn't be forgotten, and its profound and important implications ought to be remembered forever. Such a realization also totally represents and reflects the genuine voice from the bottom of my heart.)

All right, up to here, following this direction—that is, from the perspective of the highly important Planck constant, we have clearly seen the revealed fact that electron-photon work-energy relationship is the mechanism and essence of Planck constant, and viewed the two highlights coming from such a fact, also reviewed and analyzed a closely related opinion of Planck. This revealed fact, along with the specific help of these two highlights, together with the noticeable inspiration or reminder from this opinion of Planck, can play a substantial and explicit role in helping one firmly grasp the *certainty* mechanism behind the *uncertainty* principle from different angles, because this certainty mechanism is also electron-photon work-energy relationship, as presented earlier in this chapter. (At this moment, some readers may suddenly, but clearly and definitely, perceive: "Wow! The mechanism and essence of Planck constant turns out to be exactly the same thing as the *certainty* mechanism behind the *uncertainty* principle. Both are electron-photon work-energy relationship, the most essential link between an electron and its emitting photons." This perception is not only definitely correct, but also a clear reflection that these readers have a comprehensive and connected method of thinking, a crucially important and excellent quality in science. So all the readers having such a perception should have earnest congratulations on themselves! Please also accept my sincere congratulations on them.)

As a result—as a wonderful and real result, from now on, whenever people see the Planck constant h in the mathematical expressions of the uncertainties of the uncertainty principle, the first thing coming into their minds should be: this h represents the *certainty* mechanism behind the *uncertainty* principle. From now on, Planck constant h is no longer an ordinary constant; it is the *certainty* mechanism behind the *uncertainty* principle instead. From now on, the Planck constant h (which appears in either of the two mathematical expressions of the uncertainties of the uncertainty principle) begins to display its profound and revolutionary implications, begins to activate its unprecedented functions, and begins to manifest its brand-new, far-reaching and great significance!

All in all, the various discussions above have accomplished two important tasks. One was the introduction of revealing and determining the *certainty* mechanism behind the *uncertainty* principle. Another was a series of related suggestions or methods on how to chew over, digest and comprehend this certainty mechanism effectively. Through the windows of these discussions, one has not only witnessed the solid, spectacular *certainty* mechanism behind the famous *uncertainty* principle, but also viewed the magnificent and marvelous charms of this certainty mechanism from different beautiful perspectives. And through these bright windows, this spectacular *certainty* mechanism has unfolded and is spreading out its splendid charm, its great charm, and its wonderful charm to you, to me, and to all of us!

After these wonderful and memorable experiences, the *certainty* mechanism behind the *uncertainty* principle not only stands firmly and majestically in the natural world, but will also deeply camp into the minds of people. (Commentator: since the establishment of the *uncertainty* principle in the early 20th century has been said as a great achievement in science, it seems not inappropriate if one concludes that the discovery of the *certainty* mechanism behind this principle in the early 21st century is also a great and historic event in the development of science.) As a result, it seems to be a reasonable prediction or rational estimate that the spectacular *certainty* mechanism behind the famous *uncertainty* principle can and will shine out its magnificent brilliance to our entire humanity in the 21st century; the magnificent brilliance of this spectacular *certainty* mechanism belongs to our entire human beings!

Having viewed the spectacular *certainty* mechanism behind the famous *uncertainty* principle, what appears in front of us are the remarkable and profound implications that emerge from the revealing of this certainty mechanism. These implications, because of reflecting the fundamentally profound and significant consequences after having revealed this certainty mechanism, can be even more fascinating and exciting than the *certainty* mechanism underlying the *uncertainty* principle! These wonderful implications are waving their warm greetings to us.

Since the uncertainty principle has been regarded as a highly important principle that has profound implications for the way in which people view the world, clearly, revealing the *certainty* mechanism behind this principle can definitely bring about profound implications. Since the profound implications of the uncertainty principle, even prior to revealing the certainty mechanism behind it, could be said as wonderful and fascinating, undoubtedly, the profound implications *after* revealing and determining the *certainty* mechanism behind the *uncertainty* principle should or could be even more wonderful and fascinating. Thus, most readers perhaps don't want to skip over the great charm of these wonderful and fascinating implications. (If dear readers missed the opportunity to visit and view the splendid and brilliant charm of these remarkable implications due to my fault, I wouldn't be a competent guide in this journey; and it would be hard for me to forgive myself.) In fact, it seems that one can't avoid these profound implications once having known the *certainty* mechanism behind the *uncertainty* principle. Then what are these remarkable and profound implications?

We are about to view and discuss four remarkable and profound implications. First implication: it is the revealing of the *certainty* mechanism behind the *uncertainty* principle that firmly grasps the crux of the plain *reality* that this principle couldn't be truly understood (this plain *reality* has been clearly mentioned earlier in this chapter). This implication is the direct result of comparing two simple things. One is the present interpretation of this principle: there is no certainty mechanism behind the uncertainty principle, as pointed out at the beginning of this chapter. Another is the revealed or shown *fact* that there is the solid *certainty* mechanism behind the *uncertainty* principle; this certainty mechanism turns out to be electron-photon work-energy relationship, the most essential connection between an electron and its emitting photons. When one thinks these two things together, this implication thus becomes crystal clear: because the *fact* is that there is the *certainty* mechanism behind the *uncertainty* principle, of course no one can truly understand the interpretation saying that there is no certainty mechanism behind it. This is, by analogy, a bit like: because the truth is that there is motivation behind a murder case, certainly no one can really understand any interpretation or judgment that asserts there is no motivation at all behind this case. (In addition, this implication is also quite helpful to uncovering another closely related mystery, which is the widely admitted fact that quantum mechanics couldn't be really understood—this fact is known to be one of the most important, also generally acknowledged, attributes of quantum mechanics, simply because the uncertainty principle is the fundamental nature of quantum mechanics.)

Second implication: after revealing the *certainty* mechanism behind the *uncertainty* principle, it turns out that this certainty mechanism is totally "interpreted" or buried into various random effects by the present interpretation of this principle. As just mentioned above, the present interpretation says that there is no certainty mechanism at all behind the uncertainty principle. In fact, the current opinion about this principle is: it has no targeted purpose at all; it totally arises from random effects. Dominated by such a sort of interpretation and opinion, it is clear and undeniable that the *certainty* mechanism behind the *uncertainty* principle is completely treated or regarded as random effects; that is, this certainty mechanism, though not on purpose, is totally submerged or literally evaporated into a variety of random effects. Thus, clearly, even obviously, also plainly and irrefutably, revealing the *certainty* mechanism behind the *uncertainty* principle does have a profound and tremendous impact on the present interpretation and opinion of this principle, being an inevitable implication coming along with this revealing. (Related question and answer: clearly, the key to grasping this implication lies with a tangible realization of the explicit existence of this certainty mechanism; even merely from the perspective of the uncertainty principle itself, how to do that? Answer: viewing from the fully recognized, well-known *fact* that statistically, the average value of the uncertainties of the uncertainty principle is always equal to a certain positive number, which is very close to the value of Planck constant, as mentioned earlier; please notice that this average value would be simply equal to zero, IF this certainty mechanism did not exist.)

In order to have a better understanding of this implication, two pieces of closely related information are worthy to be clarified or pointed out. First, none of the various specified schemes or ideas trying to "explain"

THE CERTAINTY MECHANISM BEHIND THE UNCERTAINTY PRINCIPLE

the uncertainty principle (such as the wave/particle duality, observational interference, statistical interpretation, and thought experiment, etc.) has ever reflected the *certainty* mechanism behind the *uncertainty* principle at all. Moreover, none of these ideas has even intended to touch this certainty mechanism in fact; not to mention that the purpose of these ideas has never targeted at revealing this certainty mechanism at all. In other words, all these ideas actually have nothing to do with this certainty mechanism. Therefore, these various specified ideas, though highly prevalent and influential in "explaining" or "dealing with" the uncertainty principle, don't and can't change the simple *fact* that the *certainty* mechanism behind the *uncertainty* principle is utterly treated as random effects in the present interpretation of this principle. (Related question and answer: what is the inescapable aftermath or serious consequence of utterly "interpreting" this certainty mechanism into various random effects? Answer: confusing this certainty mechanism with truly random effects! This is self-evident and obvious; this is also clear and noticeable; this is indisputable and undeniable as well.) Second, revealing the *certainty* mechanism behind the *uncertainty* principle doesn't indicate any objection to the existence of really random effects, because this certainty mechanism doesn't and can't preclude any really random effects at all; that is, the existence of really random effects is completely not affected by whether this certainty mechanism is revealed or not. More explicitly thus more definitely, please notice that the *certainty* mechanism behind the *uncertainty* principle is caused neither by random effects nor by random factors at all, because this certainty mechanism turns out to be the definite and explicit relationship, which is electron-photon work-energy relationship (this newly established and verified relationship, as clearly pointed out several times in this chapter, is the most essential connection or link between an electron and its emitting photons, and is also the indispensable bridge between an electron and its emitting photons).

After seeing the two remarkable and profound implications discussed above, some, even many, readers might have clearly perceived that both implications turned out to be literally the consequences of having not known the *certainty* mechanism behind the *uncertainty* principle in the past (simply because this certainty mechanism had not been revealed previously). With this perception, it seems both rational and realistic that these readers can easily or naturally think of or ask the closely related question like: what's the most fundamental reason that this certainty mechanism had not been revealed before? Or what's the root cause of having been unable to reveal this certainty mechanism in the past? The answer to this question is the third profound implication *after* revealing this certainty mechanism.

Third implication: revealing the *certainty* mechanism behind the *uncertainty* principle also digs out, through a sharp contrast, the root cause of why this certainty mechanism could not be revealed before. Needless to say, the obvious reason (of this certainty mechanism could not be revealed before) was that electron-photon work-energy relationship had been unable to be established earlier, simply because this relationship has been proven to be this certainty mechanism. This inability, in turn, clearly indicated that the direct reason (of this certainty mechanism could not be revealed before) was that the new physical law, which is the law of an orbiting electron with periodic impulses emitting photons (OEPIEP law, for short), had not been discovered in the past, because electron-photon work-energy relationship is the direct result of an application of OEPIEP law, as pointed out earlier in this chapter. Thus, the *root cause*, or the most fundamental reason, of having been unable to reveal this certainty mechanism in the past turned out to be that the fundamental problem of *why* there are quantum states had not been solved. This is because the exact reason of OEPIEP law leading to the birth of electron-photon work-energy relationship (that is, the certainty mechanism behind the uncertainty principle) lies in the most fundamental role of this law: solving the very problem of *why* there are quantum states, by revealing the *quantum* mechanism of why and how photons, being the tiny and discrete quanta or particles of light, get their velocity c (the speed of light) from the electron that emits them. And so, revealing the certainty mechanism behind the uncertainty principle ultimately excavates the root cause of why this certainty mechanism could not be revealed previously. (Commentator: the excavation of this root cause is of fundamental necessity and profound significance, because the problem of *why* there are quantum states is obviously either the fundamentally important problem in quantum physics or the most fundamental problem in front of quantum mechanics.)

In addition, what should be noticed is that this root cause turns out to be also the deepest reason resulting in the serious inconsistency of quantum mechanics and general relativity (for the specific information about this serious inconsistency and this deepest reason, please go back to chapter one if necessary), this feature can enable one to see or realize this deepest reason more clearly (related reminder: as analyzed and pointed out earlier in this chapter, clearly knowing about this deepest reason is definitely fundamentally important, because the actual implications of this serious inconsistency are crucially and closely related to the fate and future of modern physics). And this feature can also enable one to see this root cause even more clearly thus grasp it more effectively and confidently, because this serious inconsistency is so well known that it has become very important general knowledge or common sense in physics. As a result, also

THE CERTAINTY MECHANISM BEHIND THE UNCERTAINTY PRINCIPLE

as a matter of fact, this fully recognized, well-known serious inconsistency turns out to be a clear and constant reminder of the explicit existence of this root cause, which can easily make or help one realize that the excavation of this root cause is actually, or should be, reasonably acceptable to most people, even if not everyone.

Having been aware of this root cause (which is that the fundamental problem of *why* there are quantum states had not been solved in the past, as dug out above), it seems neither unusual nor irrational that some, even many, insightful and sharp readers might not be content with merely knowing this root cause. They may further think of some deeply concerned, also closely related, questions like: why does this root cause exist? Has the existence of this root cause led to other profound or serious consequences (in addition to preventing from revealing the certainty mechanism behind the uncertainty principle)? From the very existence of this root cause, what painful experience can we learn or what bitter lesson can we draw? The rational and possible existence of these closely related questions signals the coming of the fourth profound implication *after* revealing the certainty mechanism behind the uncertainty principle, also an in-depth implication *after* having dug out the root cause of why this certainty mechanism could not be revealed in the past. Thus, we shall have a brief analysis and discussion of this implication before winding up the journey of this chapter.

Clearly, the key to knowing this implication actually lies in finding out the answer to such a concrete question: why had it been unable to solve the fundamentally important problem of *why* there are quantum states in the past (because this inability turned out to be the root cause of why the certainty mechanism behind the uncertainty principle could not be revealed in the past, as just seen above)? Quite obviously, the answer to this concrete question, in turn, hinges on getting the explicit answer to the following specific question like: which theory should have solved this problem? Or which theory was supposed to take on the job of tackling this fundamentally important problem?

Undoubtedly, the theory of quantum mechanics should have taken on this job, being clearly and objectively determined by the very name of this theory and its task (this task has been mentioned earlier in this chapter), as well as the fundamentally important status or position of this theory in physics. Unfortunately, the plain or "harsh" reality is that quantum mechanics does not have the ability to take on this job at all, being sufficiently evidenced or intrinsically determined by the two most important features of quantum mechanics.

The most prominent characteristic feature of quantum mechanics, also its first most important feature, is that the so-called quantum mechanics is *wave* mechanics, as clearly pointed out earlier in this chapter. In addition, what should be mentioned or noticed is: none of the knowledgeable or qualified professional people in physics want to deny the fact that quantum mechanics is *wave* mechanics, simply because they know this fact very well. (Commentator: how can *wave* mechanics solve the problem of *why* there are *quantum* states?!? It is obviously impossible; and it is dynamically thus essentially impossible! This is because *wave* mechanics doesn't and can't have the function or ability to solve the problem of *why* there are *quantum* states at all; this is because *wave* mechanics never has the magic power to possess the function or ability to solve the problem of *why* there are *quantum* states at all. How can the problem of *why* there are *quantum* states be solved with *wave* mechanics?!? It is definitely impossible; and it is clearly and unmistakably impossible! This is because *wave* mechanics has neither the function nor the ability to solve the problem of *why* there are *quantum* states—either in essence or in truth or in both.) So this most prominent characteristic feature has no choice but to tell us: it's not a surprise at all that quantum mechanics does not have the ability to solve the fundamental problem of why there are quantum states; in fact, it turns out to be rather reasonable thus quite understandable that quantum mechanics does not have such an ability. Consequently, also plainly and undeniably, in the terminology of *quantum mechanics*, the word "mechanics" is utterly unable to fulfill the task required by the word "quantum" at all. Thus, the so-called "quantum mechanics" turns out to be a purely empty terminology; it turns out that quantum mechanics factually fails to match its own label. (Related clarification or reminder: as clearly mentioned earlier in this book, being important general knowledge or common sense in science, the task or purpose of experimental tests is not to answer the questions about the *whys* or solve the problems of the *whys*, because experimental tests themselves have neither the function nor the ability to do that; instead, the task of answering these questions or solving these problems is, or is supposed to be, responsible by theories. And so, this general knowledge or common sense explicitly and unavoidably tells us that all the experimental tests of quantum mechanics don't and can't change the *reality* that quantum mechanics is unable to solve the fundamental problem of *why* there are quantum states.)

The second most important feature of quantum mechanics is the inherent nature of its core concept that is a fundamentally indispensable and crucially important concept or idea in quantum mechanics. This core concept appears as an extremely, even decisively, important core *assumption* that was fundamentally and crucially indispensable for the birth of quantum mechanics. This *assumption* says, actually claims, that all forms of energy out of electrons are released in the way of discrete units or bundles called quanta, which are known as photons nowadays. The existence of this core *assumption*, however, is actually a clear indication of such a plain *truth:* while quantum mechanics recognizes even emphasizes that photons are emitted from electrons, it is, in fact, unable to solve the problem of why and how photons are emitted from electrons at all. More specifically thus further explicitly, quantum mechanics doesn't and can't have the ability to solve the problem of why and how photons, being the tiny and discrete quanta or particles of light, get their velocity c (the speed of light) from the electron that emits them. What should be noticed is that this problem is exactly the fundamental problem of *why* there are quantum states. So this plain *truth* simply becomes that quantum mechanics is incapable of solving the fundamental problem of *why* there are quantum states. (Narrator: one can clearly and easily grasp this plain *truth* if he or she thinks it over conversely like the following. If quantum mechanics had been able to solve this fundamental problem, its fundamentally and crucially indispensable core *assumption* just mentioned above would be completely unnecessary; thus this core *assumption* would never have appeared at all. And so, the hard *fact* that quantum mechanics indispensably and desperately necessitates this core *assumption* is exactly a clear indication, or actually admits, that quantum mechanics is really unable to solve this fundamental problem, believe it or not; this hard *fact* also makes one unable to deny this inability of quantum mechanics. Of course, this inability of quantum mechanics will manifest itself even further thus become more noticeable if one thinks of or asks the simple and clear question like: if quantum mechanics had been able to solve this fundamental problem, who would have looked for trouble by proposing such a totally redundant *assumption?!?)* All in all, this second most important feature of quantum mechanics has no alternative but to tell us: it turns out to be pretty reasonable and fair that quantum mechanics doesn't have the ability to solve the fundamental problem of *why* there are quantum states.

The above two pieces of explicit and sufficient evidence, being sufficiently evidenced or intrinsically determined by the two most important features of quantum mechanics, have no choice but to show us such a clear and unavoidable conclusion: really and surely, quantum mechanics is utterly unable to solve the fundamental problem of *why* there are quantum states. (Narrator as a friendly reminder: yes, that is definitely true! This inability of quantum mechanics, as pointed out earlier in this chapter and as mentioned several times in this book, has been perplexing many brilliant physicists for a long time; that is, this inability, being a tacitly admitted *fact* in reality, has become a clearly realized or actually recognized *fact* in essence, thus also being an irrefutable or undeniable *fact* in truth.) (Related questions and answers: clearly enough, either of the two most important features presented above is sufficient to show that quantum mechanics is indeed unable to solve the fundamental problem of *why* there are quantum states, why have you mentioned both of them? Answer: in order to ensure dear readers could see this inability of quantum mechanics more clearly, I have taken the action of going the extra mile. Question: what is the relation between these two features, regarding this inability of quantum mechanics? Answer: the first feature discloses that quantum mechanics is really unable to solve this fundamental problem; the second feature actually admits such an inability. So these two features together can enable one to see this inability of quantum mechanics more clearly, perceive it more definitely, grasp it more explicitly, and get it more confidently.)

This inability of quantum mechanics turned out to be the root cause that this fundamental problem had not been solved in the past, if quantum mechanics has realized that the problem of *why* there are quantum states is the most fundamental problem in front of itself, thus agreed that it was supposed to take on the very task of dealing with such a problem. In other words, if quantum mechanics has this realization and agreement, then it becomes clear that we can come to such a rational and valid conclusion: the root cause of having been unable to reveal the *certainty* mechanism behind the *uncertainty* principle in the past was because quantum mechanics is incapable of solving the most fundamental problem it has to face, *why* there are quantum states. (*However, if quantum mechanics were unhappy or uncomfortable on seeing this root cause, or if quantum mechanics didn't want to admit or accept that this root cause is ascribed to itself alone, or if quantum mechanics is insightful enough, it can definitely have the rational, valid and effective argument like the following. Another explicit or noticeable reason of having been unable to reveal the *certainty* mechanism behind the *uncertainty* principle in the past was due to that the new physical law, which is the law of an orbiting electron with periodic impulses emitting photons, had not been discovered earlier, because this law is the *first* and *only* physical law or theory, *in the history of science*, that has the ability to solve the problem of *why* there are quantum states, by revealing the *quantum* mechanism of why and how photons, being the tiny and discrete quanta or particles of light, get their velocity c—the speed of light, from the electron that emits them. This new physical law is ready and happy to admit and accept such an explicit or noticeable reason.)

More extensively, when viewed from a large perspective or wide angle, the existence of this root cause has brought about other profound or serious consequences. These consequences can be seen more clearly when they are viewed together with a closely related clear reality or solid fact, which is that quantum mechanics and general relativity cannot both be correct because they are known to be inconsistent with each other, as explicitly pointed out by Hawking in his *A Brief History of Time* (1998 version, P.

THE CERTAINTY MECHANISM BEHIND THE UNCERTAINTY PRINCIPLE

12). In front of such a clear reality or solid fact, it seems that we cannot afford not to inspect whether quantum mechanics has the ability to solve its facing most fundamental problem—*why* there are quantum states! However, the two pieces of explicit and sufficient evidence just presented above have clearly and unavoidably shown us that quantum mechanics is utterly unable to solve this fundamental problem at all.

More than that, this inability of quantum mechanics, as analyzed and uncovered in chapter one, turns out to be also the deepest reason leading to the serious inconsistency of quantum mechanics and general relativity, which can enable one to perceive and realize this deepest reason more clearly and definitely. (Related reminder: as mentioned earlier in this chapter, this serious inconsistency actually shakes the seemingly glorious mansion of modern physics, or at least shows that this mansion cannot be stable, because quantum mechanics and general relativity are the two main theoretical pillars of modern physics. In fact, it is quite safe to say that neither knowledgeable physical experts nor competent physicists want to deny this serious inconsistency, because they know it very well. In reality, it is rather rational to conclude that all insightful physicists have clearly realized that this serious inconsistency is the deepest disaster of modern physics, also one of the most important, long-term unsolved problems in science, because they are very familiar with the fundamentally and crucially important status of quantum mechanics and general relativity in modern physics and in science. In effect, this serious inconsistency has become one of the most prominent and terrifying obstacles to the further development of physics or science.) Therefore, this serious inconsistency, via its profound and extensive, also unavoidable, consequences, has clearly demonstrated that the existence of this root cause has actually brought about other serious, even disastrous, consequences, in addition to preventing from revealing the certainty mechanism behind the uncertainty principle. Such a clear demonstration, in turn, shows or points to the fundamental importance and necessity of digging out this root cause, and is also a noticeable manifestation of the great significance of this profound implication, one of the four remarkable and profound implications after revealing the certainty mechanism behind the uncertainty principle.

Finally, what come along with this implication are two closely related questions that arise from the long-term existence of this root cause. They are: what painful experience can we learn from this root cause? What bitter lesson can we draw from this root cause? (To these questions, different people may have different answers, but I am delighted to share mine with dear readers through the following three concise points. Of course, and as always, all readers are welcome to have comments on my answer.) Point one, it seems clear, even obvious, also self-evident and irrefutable, that any theory in science, of course quantum mechanics not being an exception, as long as it is unable to solve the most fundamental problem in front of itself, cannot be said or regarded as a fundamental theory. (Impartial, rational and sensible commentator: if a theory being unable to solve the most fundamental problem in front of itself is claimed or regarded as a fundamental theory, this most fundamental problem will definitely say NO; the most fundamental nature of science will clearly say NO; the rational people in the world will impartially say NO. As a matter of fact, also as a self-evident truth, it is neither correct nor true, neither rational nor valid if anyone asserts a theory, which is unable to solve the most fundamental problem in front of itself, as a fundamental theory, because such an assertion is not, and cannot be, the fact at all.) Quite obviously, also rather rationally, this point can urge or encourage scientists to find and establish fundamental theories that target at solving the most fundamental problems in front of them. Point two, if one concludes that any theory in science, including quantum mechanics of course, as long as it is really unable to solve the most fundamental problem in front of itself, cannot be a fundamentally correct theory—either in essence or in truth or in both, then such a conclusion is not only rational and objective, but also valid and solid. In fact, this point is so obviously self-evident that it is quite obvious or pretty noticeable to most people, even if not all people. Point three, the disastrous, even catastrophic, consequences from the existence of this root cause teach us such a bitter lesson or a painful experience: there are no shortcuts in science; any theory shall not detour or skip over the most fundamental problem in front of itself; any theory should solve the most fundamental problem it has to face! If a theory detours or skips over the most fundamental problem in front of itself, the theory may subsequently bring about some extremely serious consequences or troubles.

All right, all in all, through and along the journey of this chapter, we have not only witnessed the splendid and brilliant charms of the spectacular *certainty* mechanism behind the famous *uncertainty* principle from different angles, depths and heights, but also viewed the wonderful fascinations of the profound and remarkable implications that come along with the revealing of this certainty mechanism. I do hope most readers, at least some readers, would have felt that this journey is an enjoyable experience. If my hope is true, I am really pleased too, which is also the best reward to me, the guide in this journey.

CHAPTER 7

WHY DIFFICULT TO UNDERSTAND "THE ARROW OF TIME"

[The Window on This Chapter]

Clearly enough, unless one could really understand the concept of the cosmological arrow of time, it is not possible that he or she could truly understand the chapter "The Arrow of Time" in *A Brief History of Time*, because this concept is a widely pervasive, crucially important subject there.

Where does this concept come from? It comes from the no boundary proposal. Then where does this proposal originate from?

This proposal originates from the assumed quantum theory of gravity that requires the combination of the two theories: general relativity and quantum mechanics. However, these two theories cannot both be correct because they are known to be inconsistent with each other, as Dr. Hawking has explicitly told us in his *A Brief History of Time* (1998 version, P. 12). Therefore, this assumed theory cannot be correct; accordingly, this proposal cannot be correct or valid.

Since the concept of the cosmological arrow of time comes from a proposal that cannot be correct or valid, it turns out to be both reasonable and fair if one feels it's difficult to understand this concept, thus "The Arrow of Time".

Can the cosmological arrow of time, even it is intermingled with the second law of thermodynamics via the so-called thermodynamic arrow of time, be able to "defend" or "justify" the no boundary proposal?

Plainly enough, the key to making sure that one could truly understand the chapter "The Arrow of Time" in *A Brief History of Time* is that he or she could really understand the concept of the cosmological arrow of time, because this concept pervaded that chapter (chapter nine, *A Brief History of Time*, 1998 version). It is also clear enough that, only if this concept is correct or valid, could one really understand it, because in science it is self-evident that rational people with reasonable thinking could not really understand a concept that turns out to be incorrect or invalid. That is, one can safely conclude that no rational people would claim that they could really understand a concept that turns out unable to be correct or valid. (Certainly, nobody can rule out the possibility that some, even many, insightful readers might have such a rational or sensible attitude: even some people claimed that they could understand a concept that turns out unable to be correct or valid, the so-called "understanding" of this sort is actually meaningless or pointless in science. This attitude seems reasonably acceptable to most people, even if not everyone.)

Then how to check whether the concept of the cosmological arrow of time is correct or not? How to inspect whether this concept is valid or not? (Narrator: these questions seem to have already become very important or quite meaningful, because some, even many, people have felt it's difficult to understand this concept. Accordingly, finding the answers to these questions will be the basic purpose of this chapter.)

Surely enough, as a necessary requirement or preparation to answer these questions, we need to be clear about two crucially important themes: where this concept comes from; and what this concept is about. First theme: this concept comes from the no boundary proposal suggested by Stephen Hawking. (This proposal, as explained by Hawking himself, was to describe the very early universe under the no boundary condition; that is, the initial state of the universe was finite in extent but had no singularity that outlined a boundary or edge.) Seeking the answers to the above questions thus directs us to scrutinize whether the no boundary proposal is correct or not, valid or not (the first task). Clearly, if this proposal turns out unable to be correct or valid, so does this concept. Second theme: this concept is about the direction of time in the expanding universe (the cosmological arrow of time is the direction of time in which the universe is expanding, as explained by Hawking in his *A Brief History of Time*). Looking for the answers to the above questions therefore leads us to inspect whether the idea of the expanding universe is correct or not, valid or not (the second task). If this idea turns out not to be correct or valid, neither does this concept. These two tasks will be carried out one by one as follows, in order to have the basic purpose of this chapter achieved.

Well, let us work on the first task now: to scrutinize whether the no boundary proposal is correct or not, valid or not. For doing that, we need to examine whether its theoretical basis is correct or not, valid or not. (If the theoretical basis of this proposal turns out unable to be correct or valid, so does this proposal.) As we have seen in chapter five, the theoretical basis of this proposal is the assumed quantum theory of gravity (the other name for this theory is simply quantum gravity theory or quantum gravity);

that is, the no boundary proposal is based on, or originates from, this assumed theory. Since this assumed theory has not existed yet, we ought to check the *prerequisite* for its existence in order to see the possibility to have it someday.

What is this *prerequisite* then? And can it exist? Let us look into these questions. Since the main purpose of the assumed quantum theory of gravity is to combine general relativity and quantum mechanics, clearly this *prerequisite* is that general relativity and quantum mechanics have to be inherently or fundamentally consistent with each other, so that the combination of them doesn't result in obviously ridiculous results. Since these two theories are seriously inconsistent with each other (please see chapter one about this serious inconsistency), this *prerequisite* thus becomes that there must be a way to solve this serious inconsistency. Nevertheless, this serious inconsistency has not been, cannot be, and will not be solved within the paradigm or stereotype of either general relativity or quantum mechanics or both, being clearly determined by the deepest reason or root cause of this serious inconsistency.

This deepest reason, as analyzed and dug out in chapter one, turns out to be: general relativity is unable to resolve the most fundamental problem in front of itself—*why* space and time are variable thus relative in a gravitational field, simply because it doesn't have the ability to reveal the mechanism behind this *why;* and quantum mechanics cannot resolve the most fundamental problem it has to face—*why* there are quantum states, simply because it is utterly unable to reveal the *quantum* mechanism of why and how photons, being the tiny and discrete quanta or particles of light, get their velocity c (the speed of light) from the electron that emits them. Therefore, the fundamental or inherent nature of this deepest reason sufficiently shows and explicitly determines that there is utterly no way to solve this serious inconsistency. Consequently, one has to face up to such a clear, undeniable and unchangeable reality: the *prerequisite* for the existence of this assumed theory doesn't exist at all!

Accordingly, the assumed quantum theory of gravity doesn't and can't exist at all, needless to say its correctness or validity. We have to therefore reach such an unavoidable conclusion: the no boundary proposal turns out to have been built upon a non-existent theory, being similar or equivalent to that a house or building doesn't have foundation. (Commentator: fortunately, this proposal is not a house or building! Otherwise, it's hard to imagine the consequences or aftermath.) So by analogy, this proposal turns out to be like castles in the air, flowers in a mirror, the moon in the water. (After seeing this unavoidable conclusion, the cosmological arrow of time begins to cry sadly: "Because I come from the no boundary proposal, it turns out that I am actually like a house having no foundation. So I am floating with winds!")

In fact, this unavoidable conclusion could be seen even more clearly against the background formed by the "harsh" *actuality* that the assumed quantum theory of gravity has to face (for more information about this harsh actuality, please also see chapter five). How harsh is this actuality? Let us review the closely related facts. Let these facts tell us!

As openly admitted and stated in many highly authoritative resources (for instance, the encyclopedia of physics; and *Science* Magazine, the issue of July 8, 2005, to be exact—this magazine is one of the most authoritative and influential publications in science, as mentioned earlier in this book), the quantum theory of gravity is one of the most difficult intellectual challenges physicists have ever faced, and is also one of the most difficult puzzles in science. Consequently, this assumed theory, though an ongoing program *for more than half a century* as the hottest research area in theoretical physics, has not produced fruit so far—that is, this assumed theory has not existed yet. Concisely, it has been fully recognized that the crux or "prime culprit" of this great challenge comes from the well-known *fact*, which is that general relativity and quantum mechanics are seriously inconsistent with each other. Moreover, because these two theories are the two main theoretical pillars of modern physics, this serious inconsistency has even been widely admitted as the deepest disaster of modern physics. (This deepest disaster has been ultimately incurred by the two fundamentally important, also inescapable, *facts*. One is that general relativity doesn't have the ability to resolve the most fundamental problem in front of itself, *why* space and time are variable thus relative in a gravitational field; another is that quantum mechanics cannot resolve the most fundamental problem it has to face, *why* there are quantum states, as just mentioned above.) In short, this harsh *actuality*, along with its closely related *facts*, can make or help one see more clearly that the assumed quantum theory of gravity has not been, cannot be, and will not be born within the paradigm of either general relativity or quantum mechanics or both.

This further strengthens the unavoidable conclusion above: indeed, the no boundary proposal turns out to have been built upon a non-existent theory. (Listen! Listen quietly! Listen attentively! The cosmological arrow of time is weeping more sadly and hopelessly: "Really, I am a house having no foundation at all. So I am floating with winds now, and I will be floating with winds forever. How sad my fate is! How miserable I am!")

After reading the discussions and conclusion above, one may argue: how about assuming or "forcing" the quantum theory of gravity can exist? Or how about pretending as if this theory has already existed? All right, let us see what would surely happen then. This theory requires the combination of general relativity and quantum mechanics. This combination, however, cannot produce a correct theory in fact, because Stephen Hawking has clearly mentioned: general relativity and quantum mechanics cannot both be correct because they are known to be inconsistent with each other (*A Brief History of Time*, 1998 version, P. 12). Therefore, clearly, even obviously, also plainly and undeniably, even the assumed or "forced" quantum theory of gravity cannot be correct in truth. (Narrator: in science, there is a self-evident, obvious and irrefutable rule, which is that any theory, as long as it requires the combination of two theories that cannot both be correct, definitely cannot be correct. And no one has the ability or power to change this irrefutable rule!)

Up to this moment, one can clearly see that the assumed quantum theory of gravity has to face such a dilemma: either cannot exist or is unable to be correct. Accordingly, this dilemma is also the dead alley of the claimed no boundary proposal: either is unable to be correct or cannot be valid (because this proposal is based on or originates from this assumed theory). Consequently—also unavoidably, the concept of the cosmological arrow of time, introduced in the book *A Brief History of Time*, has only two "choices" too: either cannot be correct or is unable to be valid (because this concept comes from the no boundary proposal, as mentioned and pointed out earlier). This is the plain conclusion that the cosmological arrow of time has to face! Therefore, it turns out to be reasonable and fair that one has had great difficulty understanding the concept of the cosmological arrow of time. In fact, it seems neither rational nor realistic to imagine that one could really understand this concept (because there is such a self-evident basic rationale or obvious rule in science: rational people with reasonable thinking could not really understand a concept that turns out unable to be correct or valid. And it seems rather rational and quite safe if one concludes that this self-evident basic rationale or obvious rule is, or can be, reasonably acceptable to most people, even if not everyone). And so, one should not have felt discouraged or disappointed for having not really understood this concept, because the fault or obstacle turns out not to be on the side of the people who have read this concept in *A Brief History of Time*.

Then what does the plain conclusion drawn above (the concept of the cosmological arrow of time turns out unable to be correct or valid) tell us? It clearly tells us such an inescapable *reality:* the cosmological arrow of time, even it is intermingled with the thermodynamic arrow of time conceived from the second law of thermodynamics (this law says that in any closed system disorder, or entropy, always increases with time, as explained by Hawking in his *A Brief History of Time*), still cannot "defend" or "justify" the no boundary proposal; not to mention that this proposal itself is either unable to be correct or cannot be valid from its root or basis, the assumed quantum theory of gravity. And this inescapable *reality* can be seen more clearly along the route we have passed up to here. The assumed quantum theory of gravity (which has been revealed or shown to be: either cannot exist or is unable to be correct) → the no boundary proposal (which turns out unable to be correct or valid, because it is based on this assumed theory) → the concept of the cosmological arrow of time (which, too, turns out to be: either cannot be correct or is unable to be valid, because it comes from the no boundary proposal).

This inescapable *reality* in turn has two important and unavoidable implications. One is that the opinion, which regards the no boundary proposal or the cosmological arrow of time from it or both as the explanation of the existence of the thermodynamic arrow of time, turns out to be purely an illusion (because this proposal turns out unable to be correct or valid; not to mention that the second law of thermodynamics doesn't need the explanation of the so-called thermodynamic arrow of time at all). Another is that the view, which says that this proposal, with the "help" of the weak anthropic principle, gives the explanation of why the thermodynamic and cosmological arrows of time point to the same direction, turns out to be merely a delusion of incorrect or invalid comparison.

Now, let us go to the second task: to inspect whether the idea of the expanding universe is correct or not, valid or not. In fact, this task has already been systematically completed in chapter three; and the main results obtained from this completed task have also been briefly reviewed in chapter five. (So I don't think it is necessary to repeat the specific or meticulous details here, instead it is sufficient only to present the conclusive key points extracted from the related sections or parts of those two chapters. Please go back to the detailed information in those related sections or parts, if some readers feel that these key points themselves are not tangible or specific enough.) Through the careful inspection and close scrutiny carried out in chapter three, we have gotten the explicit conclusion that the idea of the expanding universe turns out unable to be correct or valid. And this conclusion is sufficiently shown thus clearly determined by the three pieces of hard evidence, actually three hard facts. Let us concisely review them as follows.

The first evidence is that the theoretical basis of this idea, general relativity, turns out to be incorrect, for two unavoidable *realities*. One is that general relativity is unable to solve the most fundamental problem in front of itself: *why* space and time are variable thus relative in a gravitational field, simply because it is incapable of revealing the mechanism behind this *why* at all. Another is that general relativity is crucially based on special relativity. However, special relativity turns out unable to be correct, because it turns out to be clearly and seriously self-contradictory, which is because its two core or key concepts (length contraction and time dilation, respectively for interpreting length becomes shorter and time runs slower that appear in the situation of high speed) turn out to be clearly and seriously self-contradictory (this is because length contraction and time dilation are factually incompatible thus essentially contradictory with each other—they would directly deny each other if met together, as demonstrated, analyzed and concluded in chapter two; for the detailed information about this demonstration, analysis and conclusion, please see chapter two). (Commentator: if some people, especially some of those respected professional people in physics, are still psychologically reluctant towards the revealed fact that general relativity turns out to be incorrect, please think of and think about the explicit *fact* that Dr. Hawking has clearly mentioned in his *A Brief History of Time:* general relativity and quantum mechanics cannot both be correct because they are known to be inconsistent with each other—1998 version, P. 12. Undoubtedly, the irrefutable existence of this explicit fact can effectively remove or substantially reduce one's psychological reluctance in facing this revealed fact. Such an effect, needless to say, can considerably help one face, digest and realize this revealed fact calmly and rationally, particularly after this fact has been clearly shown by sufficient hard evidence, actually by more than sufficient hard evidence.)

The second evidence is that the measured microwave radiation turns out unable to be the valid evidence for the idea of the expanding universe. As analyzed and concluded in chapter three, the *prerequisite* for the measured microwave radiation to be the valid evidence for this idea is that the proposed big bang was the *only* source of the microwave radiation measured these days. That is, the real implication of this prerequisite is that in the numerous, virtually or almost incalculable, galaxies of the vast universe, no celestial bodies can be the sources (or can be the significant sources) of the microwave radiation at present. Such a prerequisite, however, cannot be valid at all in fact, simply because there is utterly no way to ensure its validity in truth. For instance, the observed features of gamma ray bursts have uncovered or clearly shown such a plain fact: in the vast universe, there are *numerous* celestial bodies that can be the sources of gamma ray bursts at present, and these celestial bodies can also be the important or significant sources of the microwave radiation measured today. Therefore, this plain fact directly invalidates this prerequisite. This invalidation has no choice but to tell us such a plain truth: it turns out that the measured microwave radiation cannot be the valid evidence for this idea.

The third evidence is that the Doppler effect turns out unable to provide the valid observational evidence for the idea of the expanding universe. As analyzed and pointed out in chapter three, the ideal prerequisite for the

Doppler effect to be able to provide the valid observational evidence for this idea is to ensure that the entire redshift of the light radiated from the observed stars is due to they are moving away from the earth. Nevertheless, because the Doppler effect on its own is unable to tell apart whether a redshift of light is due to a star moving away from the earth or due to a gravitational field, this prerequisite thus becomes that the redshift due to gravitational fields (which is gravitational redshift) is so small that it can be safely negligible, compared to the redshift due to the star moving away from the earth. That is, this *prerequisite* simply becomes that gravitational redshift must be insignificant. Such a prerequisite, however, turns out unable to hold at all, because there are at least four known factors, actually four known facts revealed by the observational results accumulated over the last several decades, that have sufficiently demonstrated thus clearly determined that this prerequisite cannot be valid (the detailed information about these known factors or facts has been presented in chapter three, as the third important thing to do in checking on whether the idea of the expanding universe could really be understood; and the key points of this detailed information have also been concisely mentioned in the middle of chapter five). Accordingly, this idea cannot be valid in truth.

All in all, each or any of the three pieces of hard evidence reviewed above is enough to show such an explicit conclusion: the idea of *the expanding universe* turns out unable to be correct or valid. Of course, this explicit conclusion is substantially consolidated by these three pieces of hard evidence together, thus becoming an explicit and solid conclusion. This explicit and solid conclusion has no alternative but to tell us such a clear and unavoidable reality: the concept of the cosmological arrow of time turns out unable to be correct or valid either, because this concept is about the direction of time in *the expanding universe*, as clearly mentioned and pointed out earlier in this chapter.

This clear and unavoidable reality further worsens the tragic fate of the cosmological arrow of time! (Listen! Listen up! Listen quietly and attentively! If listening carefully, one may be hearing the despairing voice of sobbing and crying from the cosmological arrow of time. "You see, I am a house having no foundation, so I have to be floating with winds. I have been anxiously and fearfully longing for my foundation. Now, all my walls have utterly shed off, yet the waited foundation has never appeared. Am I still a house? How sad I am! How miserable my fate is! I am totally hopeless now....") (Narrator: this heartbreaking voice, like a piece of slow and sorrowful funeral music, is pouring out the uncontrollable grief and despair of the cosmological arrow of time. This heartbreaking voice actually tells us the extremely disastrous fate of the cosmological arrow of time! This heartbreaking voice is also a plain manifestation of the inescapable, harsh reality that the cosmological arrow of time has to face. Moreover, this heartbreaking voice is the heartfelt call that can arouse our great sympathy to the cosmological arrow of time.)

In summary, it is actually impossible or unrealistic to imagine that one could really understand the concept of the cosmological arrow of time described in *A Brief History of Time*, because there are two impassable and unavoidable great obstacles to such an understanding. One is that this concept turns out unable to be correct or valid, because it comes from the no boundary proposal that has been revealed or shown to be either incorrect or invalid. Another is that this concept is about the direction of time in *the expanding universe*, whereas it turns out that the very idea of *the expanding universe* cannot be correct or valid. (Commentator: consequently, the so-called cosmological arrow of time turns out to be an incorrect or invalid concept assigned to describe an incorrect or invalid idea.) Therefore, the take home message of this chapter is: it turns out to be rather rational thus quite understandable that one could not really understand the chapter "The Arrow of Time" in *A Brief History of Time*, because this concept spread through that entire chapter. As a result—as a noticeable and pleasant result in fact, the actual effect of this take home message is: it turned out that neither disappointment nor discouragement should have been lingering in the minds of the people who could not really understand the cosmological arrow of time, thus "The Arrow of Time". Such an effect is the ultimate goal of this chapter!

CHAPTER 8

WHY DIFFICULT TO UNDERSTAND "WORMHOLES AND TIME TRAVEL"

[The Window on This Chapter]

Wow! It seems wonderfully fascinating and remarkably attractive when people think of the prospects for time travel through wormholes in the future.

When you are making a plan to drive for a journey, two things are of concern to you. One is to make sure there is a road to your destination; another is to ensure the road is passable.

Time travel (through wormholes) also needs two indispensable elements. One is the existence of wormholes that are based on *general relativity;* this is like there must be a road to the destination in your journey. Another is that these wormholes must be surely passable, which needs a region of space-time with negative energy density, interpreted with the uncertainty principle of *quantum mechanics;* this element is similar to the road to your destination must be passable.

However, Dr. Hawking has also clearly mentioned such a solid fact: *general relativity* and *quantum mechanics* cannot both be correct because they are known to be inconsistent with each other (*A Brief History of Time*, 1998 version, P. 12).

Consequently, also unavoidably, such a big question appears in front of us: can the idea of time travel (through wormholes) be correct? If this idea turns out unable to be correct, who could really understand it?!

Imagine that someday you may be traveling into the distant past and talking with your great-great-great grandfather. Envision that you can be traveling into the future more than five hundred years from now, and knowing about a lot of things that others cannot know today. How beautiful the picture is! How wonderful the beautiful picture is! How fascinating the wonderful picture is! Some, even many, people might hope that such a kind of imagination or envisioning would become true or realistic, the sooner the better, or at least with some feasible features.

This kind of imagination or envisioning is said to be as time travel or travel in time, which explores the possibility or chance about traveling into the past or the future. When people are longing for the marvelous prospects for time travel, it seems that they cannot help facing or thinking of such an important, also inescapable, question like: is time travel really feasible? Or does time travel at least have some feasible features? While the prospects for time travel are full of mysteries—at least they are now, the approach of looking for the answer to the question of this sort, as you will see, is not mysterious at all; it is simple and straightforward instead, thus could be easily understood. And the clear answer to this sort of question will be brought to light as the specific goal of this chapter.

When you are planning to drive for a journey, two things are certainly necessary. One is to make sure that there is a road to your destination; another is to ensure that the road must be passable (without obstruction or having a good maintenance, for example). Similarly, for time travel through wormholes to become feasible, two requirements are definitely indispensable. One is to make sure that wormholes really exist—that is, they should come from a correct or valid theory, a bit like there is a road to the destination in your planned journey. Another is to ensure that these wormholes are surely passable, which is similar to that the road to your destination must be passable. Specifically, this requirement necessitates a region of space-time with negative energy density, as explained by Hawking in his *A Brief History of Time;* that is, this requirement simply becomes that the interpretation of such a region ought to be valid or reliable. So if these two requirements can be satisfied, one can reach the conclusion that time travel through wormholes is feasible or with some feasible features; otherwise, one cannot come to such a conclusion. Therefore, in order to accomplish the specific goal of this chapter, what we shall do is to examine whether these two requirements can be met or not, being the two clear and definite tasks to be carried out as follows.

Now let us inspect whether the first requirement for time travel through wormholes can be satisfied or not. That is, looking for the answer to the big question: does the idea of wormholes come from a correct or valid theory? This idea originates from or is based on general relativity. This is because Einstein and Nathan Rosen, as early as in 1935, showed in a joint paper that general relativity allowed what they called "bridges," which are now referred to as wormholes. Accordingly, seeking the answer to this big question directs us to check whether the theory of general relativity is correct or not. If this theory turns out unable to be correct, so does the idea of wormholes; so does time travel through wormholes, needless to say its feasibility. (Just in case that some dear readers may still be surprised by the action to inspect whether general relativity is correct or not, please allow me to mention the solid *fact* that Hawking has clearly pointed out in his *A Brief History of Time:* general relativity and quantum mechanics cannot both be correct because they are known to be inconsistent with each other—1998 version, P. 12. Quite obviously, also rather rationally, the explicit existence of this solid fact can substantially alleviate one's surprise towards this action, which can considerably help one calmly and rationally face the conclusion obtained after finishing this action.)

Luckily enough, the task of inspecting whether general relativity is correct or not has already been completed in detail in chapter three. (The main/key points of the conclusion obtained from this completed task have also been concisely mentioned in chapters four and five. So here it seems enough that only the related key points in this conclusion will be briefly reviewed. If some readers feel these key points are not specific or tangible enough, please go back to the detailed information in the related sections or parts of those chapters, especially chapter three.) Through the detailed analysis and discussion in chapter three, we have evidently seen the clear fact that general relativity turns out unable to be correct. (Commentator: if some people are not psychologically happy or ready to face this clear fact, please think of and think over such a solid fact that has been clearly mentioned in *A Brief History of Time:* general relativity and quantum mechanics cannot both be correct because of their inconsistency with each other. Please also notice that this clear fact is actually a specific confirmation or plain manifestation of this solid fact, which can effectively reduce one's psychological obstacle in facing this clear fact.) And this clear fact is sufficiently demonstrated and determined by either of the following two unavoidable, also undeniable, *realities*. Let us have a brief review of these two realities.

One reality is such an inability: general relativity is unable to solve the most fundamental problem in front of itself, *why* space and time are variable thus relative in a gravitational field. For instance, while general relativity tells us that time runs slower in a gravitational field, it is unable to tell us *why* time runs slower in such a situation, simply because it is incapable of revealing the mechanism behind this *why*. In fact, this inability has been perplexing many insightful and brilliant physicists for a long time; that is, this inability is a tacitly admitted *fact* in reality; thus this inability is also an indisputable or undeniable *fact* in truth. (Related reminder: more specifically to be further manifest thus more noticeable, as mentioned earlier in this book, one could clearly perceive and explicitly realize, even merely from the perspective of general relativity itself, the *fact* that general relativity is really unable to solve the most fundamental problem in front of itself—*why* space and time are variable thus relative in a gravitational field, if he or she views or thinks over this *fact* in the simple and clear way like the following. If general relativity had been able to solve this most fundamental problem, its fundamentally indispensable postulate of 'invariant scales of length and time', which says that the scales of length and time at different points over an entire gravitational field are the same, would not have been necessary at all. And so, the irrefutable or undeniable *actuality* that general relativity indispensably and desperately necessitates such a postulate can enable one to perceive this *fact* clearly and realize it explicitly; of course, this irrefutable or undeniable *actuality* also makes one unable to deny this *fact*.) (Related clarification for avoiding possible or potential confusion or for eliminating the perplexity liable to occur: all the observational or experimental tests of general relativity don't and can't change the *fact* that general relativity is really unable to solve the most fundamental problem in front of itself, *why* space and time are variable thus relative in a gravitational field. This is because there is such general knowledge or common sense in science: observational or experimental tests themselves have neither the function nor the ability to answer the questions about the *whys* or solve the problems of the *whys*, so the task or purpose of observational or experimental tests is not to deal with these questions or these problems at all; instead, the task of answering these questions or solving these problems is, or is supposed to be, responsible by theories.) More directly thus more explicitly, this inability of general relativity has become even further clear and definite thereby much more noticeable, by the sharp contrast against the new theory that has solved this most fundamental problem (this new theory has been concisely introduced in the later part of chapter three). Moreover, this inability turns out to be also the deepest reason or root cause of the serious inconsistency between general relativity and quantum mechanics, as analyzed and dug out in chapter one, which indicates that this serious inconsistency, being a famous and inescapable *fact* in physics—this *fact* is so well known that it has become very important general knowledge or common sense in physics, is actually an impartial witness or constant reminder of this inability. Such a positive indication, needless to say, can substantially help one face this inability calmly and rationally.

Another reality is that general relativity is crucially based on special relativity. Nevertheless, special relativity turns out unable to be correct, because it turns out to be clearly and seriously self-contradictory, which is because its two core or key concepts (length contraction and time dilation, respectively for interpreting length becomes shorter and time runs slower that appear in the situation of high speed) turn out to be clearly and seriously self-contradictory (this is because length contraction and time dilation are factually incompatible thus essentially contradictory with each other—they would directly deny each other if met together, as demonstrated, analyzed and concluded in chapter two; for the specific and detailed information about this demonstration, analysis and conclusion, please see chapter two). Accordingly, general relativity cannot be correct too. (Or more explicitly thus more directly, as analyzed and discussed in chapter three: general relativity is *crucially* based on special relativity, whereas special relativity turns out unable to be correct because it is clearly and seriously self-contradictory; thus general relativity cannot be correct either.)

These two realities have evidently shown such a clear fact: general relativity turns out unable to be correct. This clear fact has no choice but to reveal and determine such an explicit conclusion: it turns out that the idea of wormholes introduced in *A Brief History of Time* cannot be correct, because this idea originates from or is based on general relativity, as clearly pointed out above (that is to say, the first requirement for time travel through wormholes cannot be satisfied). Consequently, also unavoidably and undeniably, the suggestion about time travel through wormholes cannot be correct either, needless to say its feasibility. As a result, it turns out to be quite reasonable thus very understandable that one has encountered great, even insurmountable, difficulties in understanding such a suggestion, because there is such a self-evident rationale or pretty obvious rule in science: it is quite normal or rather logical that rational people with reasonable thinking could not really understand a suggestion or a proposal that turns out to be incorrect. (If some people are psychologically reluctant or uncomfortable to face the plain truth that the suggestion about time travel through wormholes turns out to be incorrect, they can choose a mild expression for sounding less harsh, such as that it's indeed extremely difficult to imagine that one could really understand this suggestion because it comes from or is based on a theory that turns out unable to be correct.)

The idea of wormholes coming from a theory that turns out unable to be correct is actually (or at least in a large sense) equivalent to that there is no road to the destination in your planned journey. Not only that, this equivalence becomes further clear against the following sharp contrast. The new theory (it has been concisely introduced in the later part of chapter three), which has solved the fundamental problem of *why* space and time are variable thus relative in a gravitational field, shows that there are no wormholes at all; whereas the so-called wormholes are the product of general relativity that doesn't have the ability to solve *this* fundamental problem at all. In other words, the very reality just mentioned above, which is that general relativity is unable to solve the most fundamental problem in front of itself—*why* space and time are variable thus relative in a gravitational field, turns out to be a good or marked reminder of this equivalence. Undoubtedly, such a noticeable reminder can considerably help one see or sense this equivalence more clearly. (After seeing the information in this paragraph, it seems neither irrational nor unrealistic if some, even many, insightful and sharp readers have such a conjecture or reasoning: the so-called wormholes may be merely some kind of "compensation" for the outcome or consequence that general relativity doesn't have the ability to solve the most fundamental problem in front of itself, *why* space and time are variable thus relative in a gravitational field. The response from me: this sort of conjecture or reasoning can definitely help these readers grasp this equivalence more effectively, though the topic about this conjecture or reasoning is beyond the range of this book. So, all the readers who have gotten this type of conjecture or reasoning should be happy for themselves; and I am happy for them too.)

With this equivalence, it seems that there is no need to discuss whether the second requirement for time travel through wormholes can be met or not. Why? The reason is quite simple and obvious: even if this requirement could be met, time travel through wormholes would still not be feasible. However, as an action of going the extra mile on the topic of this chapter, I think it's better for us to be aware of the real situation about this requirement, which will be realized by carrying out the task of checking whether this requirement can be met or not. (This requirement, as mentioned above, is that wormholes must be passable, which necessitates a region of space-time with negative energy density, so this requirement actually becomes that the interpretation of such a region must be valid or reliable. More explicitly thus more directly, this requirement literally becomes that the concept of negative energy density must be valid or reliable, or the interpretation of such a concept must be valid or reliable.) Thus, if the interpretation of the concept of negative energy density is valid or reliable, then it's all right if one comes to the conclusion that this requirement can be met. On the contrary, if the interpretation of this concept turns out unable to be valid or reliable, then one has no alternative but to draw the unavoidable, also inevitable, conclusion that this requirement cannot be met.

Since the uncertainty principle is responsible for interpreting the concept of negative energy density, the task of checking whether this requirement can be met or not thus turns into checking up on whether this principle can give the valid or reliable interpretation to such a concept. (Commentator: if some people are doubtful or unsure about the necessity or importance of carrying out this task, please think of or think over the solid *fact* that has been clearly pointed out by Hawking in his *A Brief History of Time:* general relativity and quantum mechanics cannot both be correct because they are known to be inconsistent with each other—1998 version,

P. 12; along with the clear and definite reminder from another closely related *fact:* the uncertainty principle is the fundamental nature of quantum mechanics. So it is the very existence of these two closely related *facts* that points to or reminds us the necessity or importance of carrying out this task.)

About the uncertainty principle, the present explanation or opinion is: there is no certainty mechanism behind it—that is, this principle has no targeted purpose at all, and it totally arises from random effects. Thus, if such an explanation is true, it's all right to say that this principle is able to provide the valid or reliable interpretation for the concept of negative energy density. On the other hand, if it turns out that there is the *certainty* mechanism behind the *uncertainty* principle, it becomes clear enough that we have no choice but to face such a plain *reality:* the uncertainty principle cannot provide the valid or reliable interpretation for this concept. (This is, by analogy, a bit like: because the naked truth is that there is motivation behind a murder case, clearly, even obviously, any interpretation that claims or says there is no motivation at all behind this case is incorrect, and of course, cannot be valid or reliable.)

Accordingly, the conclusion on whether the second requirement for time travel through wormholes can be met hangs on the answers to the following two essentially important questions. Is there the *certainty* mechanism behind the *uncertainty* principle? If the answer is YES, what is this certainty mechanism then? (One may curiously ask: how to show or prove that there is the certainty mechanism behind the uncertainty principle? The answer is quite simple and clear: find it out! This is a bit like: having found out the continent of North America is the full proof that there is this continent; having found out gold beneath a certain place is the hard evidence that there is gold beneath this place.)

The answers to the above two questions have appeared in chapter six. (So only the key points related to these answers are mentioned here; please revisit that chapter if one wants to see the specific and detailed information about these answers.) As presented there, the *certainty* mechanism behind the *uncertainty* principle turns out to be electron-photon work-energy relationship. This relationship shows: the amount of <u>work</u> accepted by an electron over its one impulse period is equal to the <u>energy</u> of the photon emitted by the electron in the impulse period. Therefore, it turns out that indeed, there is the *certainty* mechanism behind the *uncertainty* principle. (Commentator: what should be noticed is that the revealing of this *certainty* mechanism not only shouldn't be a surprise, but also ought to be quite reasonable or rather natural in reality, because the uncertainty principle itself has actually disclosed a clear and noticeable clue that points to the explicit existence of this *certainty* mechanism. This clue, being a prominent, also well-known, basic nature of the uncertainty principle, is the fully recognized *fact* that statistically, the average value of the uncertainties of the uncertainty principle is always equal to a certain positive number, which is practically equal to the value of Planck constant; please notice that this average value would be simply equal to zero, IF this certainty mechanism did not exist. That is, even merely from the perspective of the uncertainty principle itself, one could clearly and easily sense the real and noticeable existence of this certainty mechanism, which can greatly help him or her face this certainty mechanism calmly and rationally.)

Not only that, this *certainty* mechanism has three important and definite roles with profound implications and remarkable significance. It reveals the crucially important work-energy relationship between the work accepted by an electron and the energy of its emitting photons; it represents the most essential connection or link between an electron and its emitting photons; and it is also the indispensable and unavoidable bridge between an electron and its emitting photons. Quite obviously, all these crucial and concrete roles collectively, consistently and clearly point to the inevitable and definite existence of this *certainty* mechanism, which can enable one to perceive and realize the solid existence of this *certainty* mechanism even more clearly and definitely, thus grasp it more surely and confidently. (Related question and answer: how could one grasp these crucial and concrete roles explicitly and effectively, thus grasp the *certainty* mechanism behind the *uncertainty* principle explicitly and effectively? Or what is the key to grasping these roles tangibly and definitely, thereby grasping this *certainty* mechanism tangibly and definitely? Answer: be aware or realize that the uncertainty principle is about the electrons that emit photons, because this principle is not only the fundamental nature of quantum mechanics, but also originates from quantum mechanics; and because the task of quantum mechanics is to describe how electrons emit or radiate photons, as explicitly pointed out in chapter six based on the irrefutable *fact* about the origin of this principle.)

More specifically thus more tangibly, this *certainty* mechanism possesses two other characteristic features. One is that over the range of one complete impulse period of the electron (that emits photons), the value of this certainty mechanism is equal to that of Planck constant h. (One can perceive this characteristic feature more clearly via the fully recognized *fact* just mentioned above: statistically, the average value of the uncertainties of the uncertainty principle is practically equal to the value of Planck constant h. In addition, please be reminded that this fully recognized *fact* is a prominent, also well-known, basic nature of the uncertainty principle, as just pointed out above.) Another is that this certainty mechanism turns out to be also the mechanism and essence of Planck constant. (For the detailed and specific information about these two characteristic features, please go back to chapter six if necessary.) These two features considerably consolidate or further confirm the solid and definite existence of this certainty mechanism, because each of the two mathematical expressions of the uncertainties of the uncertainty principle contains Planck constant h. (Note: the uncertainties of the uncertainty principle have two mathematical expressions. One is the uncertainty in momentum and position; another is the uncertainty in energy and time. In either of these two expressions, there is a constant; this constant is Planck constant h.) So, with these two characteristic features kept in mind, whenever people see the Planck constant h in either of the two mathematical expressions (of the uncertainties of the uncertainty principle), they can easily and clearly identify: this h represents and is the *certainty* mechanism behind the *uncertainty* principle. With and through such a clear and definite identification, one can explicitly perceive and fully realize the solid and definite existence of this *certainty* mechanism.

Now we get prepared to have the conclusion on whether the uncertainty principle can give the valid or reliable interpretation to the concept of negative energy density (that is, whether the second requirement for time travel through wormholes can be met or not). Since the revealed *fact* is that there is the *certainty* mechanism behind the *uncertainty* principle, whereas the present explanation or opinion says that there is no certainty mechanism at all behind this principle, we have to therefore draw such an unavoidable conclusion: the uncertainty principle turns out unable to provide the valid or reliable interpretation for this concept. This unavoidable conclusion has no choice but to tell us such an inescapable *reality:* the second requirement (for time travel through wormholes) cannot be met!

Up to this moment, one can clearly see the prospects for time travel through wormholes. On one side, the arena for this sort of time travel (which is wormholes) turns out to be based on a theory (general relativity) that turns out unable to be correct—that is, the first requirement (for time travel through wormholes) cannot be satisfied at all. On the other side, the interpretation of the concept of negative energy density (which is the interpretation of this arena) turns out unable to be valid or reliable—that is, the second requirement (for time travel through wormholes) cannot be satisfied either. When both sides are considered together, such a clear and solid conclusion will immediately appear as a plain fact: *it is definitely impossible for anybody to have time travel or space travel through wormholes; there is no prospect at all for the prospects of this sort of travel!* (With such a clear and solid conclusion right in front of us, it has thus become groundless thereby meaningless to discuss whether the consistent histories approach or the alternative histories hypothesis should be obeyed

A WONDERFUL GIFT TO THE READERS OF "A BRIEF HISTORY OF TIME"

in time travel through wormholes: it turns out that the very *prerequisite* for having such a discussion doesn't exist at all!)

If some dear readers are psychologically reluctant or hesitant towards the above conclusion about the prospects for time travel through wormholes, there is a specific and clear clue that can considerably help them remove or reduce this reluctance or hesitation. Such a clue is the well-known and undeniable *fact*, which is the serious inconsistency of general relativity and quantum mechanics; along with the reminder from another closely related *fact:* the uncertainty principle is the fundamental nature of quantum mechanics; together with the very *fact* that this principle is responsible for interpreting the second indispensable requirement for time travel through wormholes. More explicitly to be more perceptible, please remember that Professor Hawking has clearly pointed out such a solid fact in his *A Brief History of Time* (1998 version, P. 12): general relativity and quantum mechanics cannot both be correct because they are known to be inconsistent with each other. Quite reasonably, also pretty obviously, the undeniable existence of this solid fact can not only effectively remove or reduce one's psychological reluctance or hesitation towards the conclusion above, but also enable one to perceive or realize: this conclusion is actually quite reasonable and fair—at least not a surprise at all. This effect, needless to say, is a substantial help for one to face this conclusion calmly, rationally and bravely. (Commentator: interestingly enough, while general relativity and quantum mechanics are seriously inconsistent with each other, their true implications for time travel through wormholes are really consistent. Both theories point to the same direction, in which both of the two requirements for time travel through wormholes cannot be satisfied: general relativity points to the first requirement; the uncertainty principle of quantum mechanics points to the second requirement.)

If some people feel disappointed towards the conclusion above for my destroying their dream of time travel through wormholes, I am really sympathetic to them. But I think that telling the truth can avoid a much bigger disappointment, because such a disappointment would be coming towards them if they still believed the feasibility or achievability of time travel through wormholes. If you have not been convinced yet, you can keep on your dream, of course. But you are welcome to invite me to become involved in your dream. If you get ready for time travel, please don't forget to tell me. I also want to know about some of the detailed information about my great-great-great grandfather; I never have his picture, for example. And when you come back from time travel through wormholes, you can get a check of one million dollars from me. This check is valid for a long time. At the moment of your leaving for time travel, I shall sign it. I am going to tell my two children (son, Bert and daughter, Briana) to keep this check; and if necessary they will in turn ask their children to do that, and so on. But my bet is that this check will be waiting there forever, from one generation to another generation....

Anyway, if you have found that "Wormholes and Time Travel" in *A Brief History of Time* looks more like a science fiction, you are not wrong. If you have felt that it's rather difficult to understand time travel through wormholes, you are not alone. If you have realized that this chapter reveals the secrets of why it's really not easy to understand "Wormholes and Time Travel", you are not alone either. This realization can enable you to be well aware of such a plain and pleasurable fact: it turns out to be pretty reasonable and fair that one couldn't really understand "Wormholes and Time Travel".

CHAPTER 9

WHY DIFFICULT TO UNDERSTAND "THE UNIFICATION OF PHYSICS"

[The Window on This Chapter]

Thousands of people were excitedly acclaiming when a long and deep railway tunnel across a big mountain was finally opened up.

More people were exultantly cheering when the different sections of an entire railway line were eventually joined up.

These spectacular occasions might be a vivid reminder of a grand dream on the great project in science: when will the historic moment of successfully connecting general relativity and quantum mechanics be coming? Can this moment be coming?

However, we have to face such a clear and unavoidable *reality:* general relativity and quantum mechanics, the two main theoretical pillars of modern physics, are known to be inconsistent with each other (*A Brief History of Time*, Dr. Hawking, 1998 version, P. 12).

With this inconsistency, what are the prospects for pursuing the unification of physics?

Under this inconsistency, was it reasonable to expect that one could really understand "The Unification of Physics" in *A Brief History of Time?*

There is a large and famous building on the earth, and now its age is nearly ninety years old. This building is so huge and splendid that it looks like a wonderful mansion from the outside. Unfortunately, however, about fifty years ago, the owner of this building noticed that its three main pillars were not compatible at all; in fact, they were seriously inconsistent with each other! So this building was facing a great danger. This owner had no choice but to hire a highly professional contractor to deal with the big and urgent trouble.

The contractor had two major techniques to fix the serious inconsistency and incompatibility of the three main pillars. One was to use a kind of material composed of numerous small and separate particles; another was to bind these three pillars together with many, many very thin ropes. But rather disappointedly, neither of them was workable; even these two techniques together still couldn't work either. Of course, this contractor also tried several other techniques, but none of them worked still. So this building is facing an even greater danger today.

With the title of this chapter as a reminder, and by what has been seen through [The Window on This Chapter], one might have perceived that these three pillars respectively represent general relativity, quantum mechanics, and the uncertainty principle from it. With this perception, it seems not difficult that one could have then realized that this large building turns out to be the epitome of modern physics. Along with this perception and realization, one might have further sensed or conjectured that this kind of material made of numerous small and separate particles stands for the gravitons *assumed* in the so-called supergravity theories, and that these many, many very thin ropes symbolize the various string theories. Yes! This perception, realization and conjecture are correct!

The terrible situation described above thus displays two harsh facts, also two big and pressing problems! One is that the serious inconsistency and incompatibility among general relativity, quantum mechanics, and its uncertainty principle had been noticed and recognized for more than half a century. Another is that the present ongoing theories, such as the various string theories and the supergravity theories, have not resolved this serious inconsistency and incompatibility yet, despite the diligent and unceasing efforts of many intelligent and capable physicists over several decades. So these two big and urgent problems are anxiously waiting for us to take quick and effective actions!

The existence of these two big and urgent problems has determined that the intended actions should achieve the following two clear and definite goals. First, for general relativity, quantum mechanics, and the uncertainty principle, the goal is to analyze the deepest reason leading to this serious inconsistency, dig out the root cause or "prime culprit" resulting in this grave incompatibility, thus find out the crux or essence of this serious inconsistency and incompatibility. That is, the first goal is to find out, along the direction of internal route, the internal cause or the fundamental reason behind this serious inconsistency and incompatibility. Second, to the present ongoing theories that have not yet tackled this serious inconsistency and incompatibility, the goal is to check on whether they are really able to solve this sort of inconsistency and incompatibility, and to inspect whether these currently 'fashionable' theories are really correct or not, truly valid or not. In other words, the second goal is to uncover, following the direction of external route, the external reasons of the inability of these ongoing theories. These two concrete goals have clearly told us what specific actions to be taken in this chapter.

In the anticipated actions towards the first goal, we shall look for the answers to the following fundamentally important questions. Can the serious inconsistency of general relativity and quantum mechanics be solved (within the paradigm or framework of these two theories)? Is it possible to work out or tackle the grave incompatibility between general relativity and the uncertainty principle? (Prompter: Oh, by the way, please don't forget to inspect whether the idea in itself of trying to intermingle the uncertainty principle with general relativity is really correct or not, truly wise or not.)

Now, let us go to the actions towards the first goal—seeking the answer to the fundamentally and crucially important question with profound implications and great significance: can the serious inconsistency between general relativity and quantum mechanics be solved (within the paradigm of either general relativity or quantum mechanics or both)? In order to do that, we must be clear about the *prerequisite* to solve this serious inconsistency.

What is this *prerequisite* then? It comes from the characteristic features of general relativity and quantum mechanics. As discussed in chapter one, general relativity is built or based on *one set* of assumptions, hypotheses and postulates (AHPs, for short), whereas quantum mechanics is built or based on *another set* of AHPs (as pointed out and listed in chapter one, general relativity hires at least five AHPs, and quantum mechanics employs at least four AHPs). So, clearly enough, knowing the relationship between these two sets of AHPs is the *prerequisite* to solve the serious inconsistency of these two theories (that is, knowing this relationship is the indispensable requirement for these two theories to be connected or combined correctly thus successfully). However, there is no way to establish the relationship between these two sets of AHPs at all, simply because there is no relationship between the set of AHPs underlying general relativity and the set of AHPs underlying quantum mechanics. Therefore, this *prerequisite* cannot be satisfied at all, being a clear and undeniable *fact*.

This clear and undeniable *fact*, in turn, has no alternative but to tell us another plain or solid *fact:* it is not possible to resolve the serious inconsistency of general relativity and quantum mechanics within the paradigm of either general relativity or quantum mechanics or both! Moreover, the actual implication of this solid *fact* is that any efforts attempting to solve this serious inconsistency have been and will be fruitless. (One might ask:

both general relativity and quantum mechanics have agreed with observational and experimental results, why is it still impossible to solve the serious inconsistency between them? The answer is: these agreements still don't and can't change the unchangeable *reality* that it is impossible to find out the relationship between the two sets of AHPs (assumptions, hypotheses and postulates) that separately underlie these two theories, whereas knowing this relationship is the *prerequisite* to solve this serious inconsistency.)

Having seen the solid *fact* above, one may be wondering: since the root cause or "prime culprit" of being unable to tackle the serious inconsistency of general relativity and quantum mechanics turns out to be due to the two unconnected sets of assumptions, hypotheses and postulates (AHPs) separately employed by these two theories, why did they have to employ these AHPs? This is indeed a wise and fundamentally important question, though I am not trying to flatter anybody. Accordingly, it is of fundamental importance to be aware of the answer to such a question. So let us find out this answer (actually we are going to review this answer, because the detailed information about this answer has already appeared in the middle of chapter one).

The answer to this question lies in the deeper features of both general relativity and quantum mechanics. As explained in chapter one, general relativity has to employ one set of AHPs (assumptions, hypotheses and postulates), because it is unable to solve the most fundamental problem in front of itself, *why* space and time are variable thus relative in a gravitational field. For instance, while general relativity tells us that time runs slower in a gravitational field, it is unable to tell us *why* time runs slower in such a situation, because it is incapable of revealing the mechanism behind this *why* (in fact, this inability has been perplexing many brilliant physicists for a long time; that is, this inability, being a tacitly admitted *fact* in truth, is actually either a clearly perceived *reality* or an explicitly realized *reality* in essence, thus also being an irrefutable or undeniable *reality* of course). More specifically thus more noticeably, one could clearly perceive and explicitly realize, even merely from a basic and prominent feature of general relativity or simply from the perspective of general relativity itself, the *fact* that general relativity is really unable to solve the most fundamental problem in front of itself—*why* space and time are variable thus relative in a gravitational field, as long as he or she views or thinks over this *fact* in the simple and clear way like the following. If general relativity had been able to solve this most fundamental problem, its fundamentally indispensable postulate of 'invariant scales of length and time', which says that the scales of length and time at different points over an entire gravitational field are the same, would not have been necessary at all.

And so, the irrefutable or undeniable *actuality* that general relativity indispensably and desperately needs such a postulate can enable one to perceive this *fact* clearly and realize it explicitly; this irrefutable or undeniable *actuality* also unavoidably shows that one has no way to deny this *fact* at all. (Related clarification or reminder: as mentioned in chapter one, being important general knowledge or common sense in science, observational or experimental tests themselves have neither the function nor the ability to answer the questions about the *whys* or solve the problems of the *whys*, thus the task or purpose of observational or experimental tests is not to deal with these questions or these problems at all; instead, the task of answering these questions or solving these problems is taken on or responsible by theories, or is supposed to be taken on or responsible by theories. So this general knowledge or common sense explicitly and unavoidably tells us that all the observational or experimental tests of general relativity don't and can't change the *fact* that general relativity is really unable to solve the most fundamental problem in front of itself, *why* space and time are variable thus relative in a gravitational field.) Moreover, this inability of general relativity becomes crystal clear thus much more noticeable, by the sharp contrast against the new theory that has solved *this* most fundamental problem (this new theory has been briefly introduced in the later part of chapter three).

Also as pointed out in chapter one, quantum mechanics has to employ another set of assumptions, hypotheses and postulates (AHPs), because it is unable to solve the most fundamental problem in front of itself, *why* there are quantum states, simply because it is incapable of revealing the mechanism behind this *why* at all. (Narrator: in fact, this inability has been baffling many excellent physicists for a very long time; that is, this inability is a clearly noticed *fact* or undeniable *reality*. The key to sensing or realizing this inability lies with: the so-called quantum mechanics is *wave* mechanics that doesn't and can't have the function or ability to solve the problem of *why* there are *quantum* states at all.) Specifically, quantum mechanics is unable to reveal the *quantum* mechanism of why and how photons, being the tiny and discrete quanta or particles of light, get their velocity c (the speed of light) from the electron that emits them. More specifically to be even more conspicuous and impressive, as mentioned earlier in this book, one could clearly perceive and explicitly realize, even merely from the large perspective or basic features of quantum mechanics itself, the basic *fact* that quantum mechanics is really unable to solve the most fundamental problem in front of itself—*why* there are quantum states, if he or she views or thinks over this *fact* in the simple and clear way as follows. If quantum mechanics had been able to solve this most fundamental problem, its core *assumption*, or the fundamentally indispensable *assumption* for the birth of quantum mechanics, which says that all forms of energy released from electrons are in the way of discrete units or bundles called quanta (which are known as photons nowadays), would not have been necessary at all; thus this core *assumption* would never have appeared. And so, the irrefutable and unavoidable naked *truth* that quantum mechanics indispensably and desperately needs this core *assumption* can enable one to perceive this basic *fact* clearly and realize it explicitly; this naked *truth* also makes one have neither reason nor excuse to deny this basic *fact*. (Related clarification or reminder: as mentioned earlier in this book, being important general knowledge or common sense in science, the task or purpose of experimental tests is not to answer the questions about the *whys* or solve the problems of the *whys*, because experimental tests themselves have neither the function nor the ability to do that; the task of answering these questions or solving these problems is undertaken, or is supposed to be undertaken, by theories instead. Thus this general knowledge or common sense clearly and explicitly, also unavoidably, tells us that all the experimental tests of quantum mechanics don't and can't change the basic *fact* that quantum mechanics is really unable to solve the most fundamental problem in front of itself—*why* there are quantum states.) Moreover, this inability of quantum mechanics can be seen more clearly after *this* most fundamental problem has been solved with a new theory (in chapter six). All in all, the reason that either of general relativity and quantum mechanics has to employ its own set of AHPs is that neither of these two theories is able to solve the most fundamental problem in front of itself.

The analysis and discussion above have revealed the two basic features about the serious inconsistency of general relativity and quantum mechanics. One is that this serious inconsistency turns out to be fundamentally inherent; that is, these two theories are inherently and permanently irreconcilable from their roots. Another is that the reason of this serious inconsistency is into the essential depth of mechanism and cause, rather than on the superficial level of phenomena and effects. These two features clearly and sufficiently demonstrate that this serious inconsistency turns out to be permanently irreparable in fact. This clear and sufficient demonstration has no choice but to tell us such an inescapable and plain *truth:* no other theories (including string theory along with its variants and supergravity theories) have the ability to solve this serious inconsistency. This very *truth* is actually equivalent to such a "harsh" *reality:* it is definitely impossible to solve the serious inconsistency of general relativity and quantum mechanics, simply because there is utterly no way to solve this serious inconsistency; that is, this serious inconsistency has not been, cannot be, and will not be solved.

WHY DIFFICULT TO UNDERSTAND "THE UNIFICATION OF PHYSICS"

This "harsh" *reality* might give rise to one's considerable concern about the grave incompatibility between general relativity and the uncertainty principle, considering this principle is the fundamental nature of quantum mechanics. Some, even many, attentive readers may logically think of or raise the closely related questions like: can this grave incompatibility face the same fate? Is it possible to solve this grave incompatibility? And what are the prospects for the various attempts to solve this grave incompatibility? (Narrator: these questions are crucially important, corresponding to the fundamentally important status of general relativity and the uncertainty principle in modern physics.)

For at least two reasons, we shall take actions to seek the answer to the crucially important question: is it possible to solve the grave incompatibility between general relativity and the uncertainty principle? One reason is from the practical viewpoint: the uncertainty principle has been mentioned most frequently only next to general relativity in the book *A Brief History of Time* (this principle has appeared in and pervaded several chapters of that book). Thus, the answer to this question can play an important role in revealing the secrets of why it's difficult to understand *A Brief History of Time*. Another reason is that Hawking has emphasized in his *A Brief History of Time:* incorporation or combination general relativity with the uncertainty principle is the indispensable first step towards the quest for the unification of physics. Accordingly, the answer to this question is also a necessary component for us to grasp the main points of this chapter.

Clearly, the answer to the big question, which is whether the grave incompatibility between general relativity and the uncertainty principle can be solved or not, depends on whether the *prerequisite* to solve this grave incompatibility can be satisfied or not. If this prerequisite could be satisfied, it would be possible to solve this grave incompatibility. On the contrary, if there is no way to meet this prerequisite at all, then one has to say, to be rational and fair, that it is definitely not possible to solve this grave incompatibility.

Then can this *prerequisite* be met? Let us check on it by analyzing the related basic features of both general relativity and the uncertainty principle. As just mentioned above, general relativity is built on *one set* of assumptions, hypotheses and postulates (AHPs, for short. As to the specific information about the five AHPs hired by general relativity, please go back to chapter one if necessary). On the other hand, the uncertainty principle is built upon *another set* of AHPs, including the assumptions of energy conservation and momentum conservation, as well as the hypothesis of probabilities. (Probability is used to describe the likelihood or chance that a particular thing or event will happen; for example, one can say that there is a 70% probability that there will be rain on the coming Friday. As to energy conservation and momentum conservation, please see the Glossary of this book if necessary.) So, clearly, even obviously, knowing the relationship between these two sets of AHPs is the *prerequisite* to solve the grave incompatibility of general relativity and the uncertainty principle—that is to say, knowing this relationship is the indispensable condition or minimum requirement for these two theories to be combined or incorporated correctly, being the necessary first step towards successfully solving this grave incompatibility. However, there is utterly no way to establish the relationship between these two sets of AHPs, simply because there is no relationship between the set of AHPs underlying general relativity and the set of AHPs underlying the uncertainty principle. Therefore, we have no choice but to reach the plain conclusion that this *prerequisite* cannot be satisfied at all (even though I feel regrettable about this conclusion in my heart). This plain conclusion sufficiently determines and demonstrates such a clear, even obvious, also unavoidable, *reality:* it is definitely not possible to solve the grave incompatibility between general relativity and the uncertainty principle.

The above *reality* might sound frustrating; the following questions, nevertheless, seem to be a consolation—at least in a sense, or when one views the *reality* above from different angles. Is it essentially necessary to combine the uncertainty principle with general relativity? Is it really wise having attempted to intermingle these two theories? Or are the various actions that have been taken to "force" these two theories together truly correct or proper?

The answers to these questions depend on whether the present interpretation of (or the current view on) the uncertainty principle is correct or not, valid or not. Specifically, if the present interpretation of this principle is correct or valid, the answer is YES. On the other hand, if this present interpretation turns out unable to be correct or valid, of course the answer has to be NO, to be rational, objective and fair. (Commentator: yes; because it is not only obvious but also self-evident that in science, a theory or principle, as long as it hasn't been given a correct or valid interpretation, shouldn't be intermingled or combined with other theories. This is a bit like that a rotten vegetable shouldn't be mixed with fresh vegetables when one prepares food.)

Then how to inspect whether the present interpretation of the uncertainty principle is correct or not, valid or not? The present interpretation or view about the uncertainty principle is: there is no certainty mechanism behind it—that is to say, this principle has no targeted purpose at all, and it entirely arises from random effects. Thus, if there is no certainty mechanism behind this principle, this present interpretation is correct or valid. But on the contrary, if there is the *certainty* mechanism behind the *uncer-*

tainty principle, then we have no choice but to face the explicit and definite, also inescapable, *fact* that the present interpretation of this principle turns out unable to be correct or valid. (This is a bit like that, because an exquisite box contains a wonderful gift, any saying that the box contains nothing cannot be correct or valid.) As a result, the conclusion on this inspection hangs on the answers to the following two essentially important questions. Is there the *certainty* mechanism behind the *uncertainty* principle? If the answer is YES, what is this certainty mechanism then?

No worry, no hurry: the task of looking for the answers to these questions has been completed in chapter six, and some key points related to these answers have also been concisely mentioned in chapter eight (thus it seems that there is no need to present too much detailed information here; only the core points in these answers are enough at this moment).

As revealed and proven in chapter six, the *certainty* mechanism behind the *uncertainty* principle turns out to be electron-photon work-energy relationship. This relationship (that is, this *certainty* mechanism) shows us: the amount of <u>work</u> accepted by an electron over its one impulse period is equal to the <u>energy</u> of the photon emitted by the electron in the impulse period. Therefore, it turns out that really and truly, there is the *certainty* mechanism behind the *uncertainty* principle; this is a bit like that, having found out the continent of North America is the hard evidence that there is this continent. (Narrator: moreover, this *certainty* mechanism has three crucial and definite functions with profound implications and great significance. It reflects the crucially important work-energy relationship between the work accepted by an electron and the energy of its emitting photons; it represents the most essential connection or link between an electron and its emitting photons; and it is also the indispensable bridge between an electron and its emitting photons. Undoubtedly, these crucial and definite functions further manifest or demonstrate the explicit existence of this *certainty* mechanism, which can substantially help one perceive this *certainty* mechanism more distinctly and impressively.) (Related question and answer: how could one grasp these crucial and definite functions tangibly and explicitly, thus grasp the *certainty* mechanism behind the *uncertainty* principle tangibly and explicitly? Answer: be aware or realize that the uncertainty principle is about the electrons that emit photons, because this principle is not only the fundamental nature of quantum mechanics, but also originates from quantum mechanics; and because the task of quantum mechanics is to describe how electrons emit or radiate photons, as explicitly pointed out in chapter six based on the clear and indisputable *fact* about the origin of this principle.)

Not only that, this *certainty* mechanism has a specific, also prominent, feature, which is that over the range of one complete impulse period of the electron (that emits photons), the value of this *certainty* mechanism is equal to that of Planck constant h. Such a specific and noticeable feature can tangibly and explicitly help one realize the clear and definite existence of this *certainty* mechanism even more distinctly and impressively, because each of the two mathematical expressions of the uncertainties of the uncertainty principle contains Planck constant h. (Note: as pointed out in chapter six, and as mentioned in chapter eight, the uncertainties of the uncertainty principle have two forms of mathematical expressions. One is the uncertainty in momentum and position; another is the uncertainty in energy and time. In either of these two expressions, there is a constant; this constant is Planck constant h.)

More specifically thus more clearly, what is closely related to this specific feature is another revealed or shown fact: electron-photon work-energy relationship has also been revealed or proven, still in chapter six, to be the mechanism and essence of Planck constant too (that is, the *certainty* mechanism behind the *uncertainty* principle turns out to be the same thing as the mechanism and essence of Planck constant). Such a revealed fact can explicitly help one perceive and recognize the solid existence of this *certainty* mechanism even further, simple because Planck constant h appears in either of the two mathematical expressions of the uncertainties of the uncertainty principle, as just pointed out above. And so, with this revealed fact kept in mind, whenever one sees the Planck constant h in either of these two mathematical expressions, it seems not difficult that he or she can clearly and definitely realize: this h represents and is the *certainty* mechanism behind the *uncertainty* principle! (Commentator: undoubtedly, such a clear and definite realization can well enable one to have a good grasp of this *certainty* mechanism.)

More directly to be more perceptible, believe it or not, even merely from the angle of the uncertainty principle itself, one can also perceive a clear clue that actually points to the explicit existence of this *certainty* mechanism. This clear clue is the fully recognized *fact* that statistically, the average value of the uncertainties of the uncertainty principle is more or less equal to the value of Planck constant h; please carefully notice that this average value would be simply equal to zero, IF this certainty mechanism did not exist. Thus, such a fully recognized *fact* turns out to be literally a clear indication or marked reminder of the solid existence of this *certainty* mechanism, which can easily enable one to realize: the revealing of this *certainty* mechanism turns out to be not only definitely reasonable but also quite natural. Undoubtedly, such a realization can effectively remove or substantially reduce one's psychological obstacle in facing this *certainty* mechanism. (Related questions and answers: having seen the (newly revealed) *certainty* mechanism behind the *uncertainty* principle, some attentive and sharp readers might think of or raise the related question like 'where does this *certainty* mechanism go according to the present interpretation of this principle?' Answer: this *certainty* mechanism, because it had not been revealed in the past, is totally "interpreted" or buried into various random effects that, by their definition, do not include any certainty mechanism at all. For instance, there are various "explanations" or descriptions about the uncertainty principle, including the wave/particle duality, observational interference, and statistical interpretation, and so on. Nevertheless, none of these "explanations" or descriptions have the ability to find out or reveal this certainty mechanism at all; not to mention that their intention has never ever been to touch on this certainty mechanism! In fact, the purpose of these "explanations" or descriptions is not to find out or reveal this certainty mechanism at all; it thus becomes quite reasonable and fair that they are unable to reveal this certainty mechanism. As a result, it is not a surprise that the *certainty* mechanism behind the *uncertainty* principle is utterly treated or viewed as purely random effects in the present interpretation of this principle. Question: what is the inescapable aftermath or inevitable consequence of utterly "interpreting" this certainty mechanism into various random effects? Answer: confusing this certainty mechanism with truly random effects! This is self-evident and obvious; this is also clear and noticeable; this is indisputable and undeniable as well.)

All in all, after revealing the *certainty* mechanism behind the *uncertainty* principle, the existence of this *certainty* mechanism is a proven fact, also a clear and undeniable fact! (Again, this is a bit like that: having found out the continent of North America is the irrefutable fact that there is this continent.) Moreover, through the pieces of hard evidence and closely related information presented above, this *certainty* mechanism demonstrates or manifests itself clearly from different directions or angles, a bit like 3-D visual effects, which can explicitly and substantially help one realize such a clear and definite fact: really and truly, it turns out that there is the *certainty* mechanism behind the *uncertainty* principle.

Now, one can clearly see the solid conclusion on this inspection. Since the clear *fact* is that there is the *certainty* mechanism behind the *uncertainty* principle, whereas the present interpretation or view says that there is no certainty mechanism at all behind this principle, we have to therefore draw such a solid and unavoidable conclusion: the present interpretation of the uncertainty principle turns out unable to be correct or valid. (This conclusion, by analogy or when viewed from the angle of comprehension, is a bit like: because the plain fact is that there is motivation behind a murder case, clearly, even obviously, any interpretation, which says that there is no motivation at all behind this case, cannot be correct or valid.)

This solid and unavoidable conclusion, in turn, clearly determines and sufficiently demonstrates that the answers to the above-mentioned questions are simply as follows. It turns out that essentially, it is not necessary to combine the uncertainty principle with general relativity. It turns out that it is not really wise having attempted to intermingle these two theories. And it turns out that the various actions having been taken to "force" these two theories together cannot be truly correct or proper.

At this moment, the owner of this large and famous building (mentioned at the beginning of this chapter) may become even more worried, because he has clearly seen the three disappointing *facts* that his cherished building has to face. First, it is really not possible to resolve the serious inconsistency of general relativity and quantum mechanics, simply because the *prerequisite* to resolve this serious inconsistency cannot be satisfied at all. Second, it is definitely impossible to solve the grave incompatibility between general relativity and the uncertainty principle either, because there is utterly no way to meet the *prerequisite* for solving this grave incompatibility too. Third, what seems to be ironically or interestingly true is that even the very decision of trying to intermingle the uncertainty principle with general relativity turns out unable to be really wise in essence; accordingly, the various actions that have been taken to "force" these two theories together cannot be truly correct or proper in fact. (Narrator: in the face of these three hard facts, this owner is undoubtedly very anxious about the perilous situation of his loved building. However, he hasn't yet lost his patience with waiting for the results from the expected actions towards the second goal mentioned at the beginning of this chapter.)

As pointed out earlier, despite the great diligence of many intelligent and capable experts over several decades, the present ongoing theories, including the various string theories and the supergravity theories, haven't yet resolved the serious inconsistency of general relativity and quantum mechanics, and the grave incompatibility between general relativity and the uncertainty principle. For that reason, the actions towards the second goal ought to complete the following two clear and definite tasks. The first task is to check whether or not these ongoing theories really have the ability to resolve this kind of inconsistency and incompatibility; the second task is to inspect whether these ongoing theories are really correct or not, truly valid or not. (In addition, some insightful experts have, in fact, suspected the feasibility of string theory and its variants, with the rational questioning like: string theory, or superstring theory, or some other variant on this theme, like the final "theory of everything", a step on the road or a blind alley? The suspicion of this kind also indicates that we should carry out these two explicit tasks.)

Let us start our work on the first task. Clearly, whether or not these ongoing theories are able to resolve this serious inconsistency and incompatibility depends on whether they can satisfy the *prerequisite* to tackle such an inconsistency and incompatibility. As discussed earlier, given that general relativity and quantum mechanics are separately based on two different sets of assumptions, hypotheses and postulates (AHPs), knowing the relationship between these two sets of AHPs is the *prerequisite* to resolve the serious inconsistency of these two theories. However, there is utterly no way to find and establish the relationship between these two sets of AHPs, simply because there is no relationship between the set of AHPs underlying general relativity and the set of AHPs underlying quantum mechanics. Consequently, one has no choice but to face the unavoidable *reality* that these ongoing theories, such as string theory and its variants like the heterotic string, cannot satisfy this *prerequisite* at all, not to mention that their target has never intended to meet this prerequisite! Accordingly, one cannot avoid such a simple *fact:* these ongoing theories are actually unable to tackle the serious inconsistency of general relativity and quantum mechanics at all. Similarly, these ongoing theories don't have the ability to resolve the grave incompatibility between general relativity and the uncertainty principle either. In other words, it turns out that these ongoing theories, though not on purpose, have been "forced" to take on the work far beyond their ability, a bit like forcing a horse to drag a train. (Narrator: now, the owner of this large building seems to have realized that most probably, he had hired an incompetent contractor—at least in a large sense. But this owner hasn't yet totally given up his residual hope; he still places his last hope on the result from carrying out the second task just mentioned above. So let us work on this task.)

For carrying out the task of inspecting whether these ongoing theories are really correct or not, truly valid or not, we ought to look closely at whether they indispensably depend on the existence of something that has been thought to exist, but doesn't exist in fact. For instance, these present ongoing theories (including string theory, along with its variants, and the supergravity theories) are indispensably dependent on the presumed or claimed existence of the so-called *gravitons*, which are the guessed small particles that are assumed to carry gravitational force. That is, if there is sufficient evidence showing that the guessed gravitons don't exist at all, one has to agree that these ongoing theories turn out unable to be really correct or valid, to be rational, objective and fair. This task has thus turned into inspecting whether the guessed gravitons really exist or not, which is to be implemented in the following two specific steps.

The first step is actually a preparation for buffering one's possible psychological shock if the gravitons don't exist indeed. Some insightful and sharp experts have already doubted the existence of the gravitons. For instance, the detection of gravitons has been raised as one of the 100 big remaining questions in *Science* Magazine (the issue of July 8, 2005, to be exact. This magazine is widely known to be one of the most authoritative and influential publications in science), because nobody has spotted the gravitons. (In fact, no one has detected the gravitons even up to now, 2017.) Along with the doubt above, please allow me to provide a clear clue as a noticeable reminder of this very doubt: the speculated existence of the guessed gravitons was originally "inspired" from the theory of quantum mechanics. However, this theory is even unable to solve the most fundamental problem in front of itself—*why* there are quantum states (simply because it is incapable of revealing the mechanism behind this *why* at all). As a matter of fact, this inability of quantum mechanics has been perplexing many brilliant physicists for a long time; that is, this inability, being a tacitly admitted *reality*, is either an explicitly realized *reality* or an actually recognized *reality*, thus also an irrefutable or undeniable *reality*, as pointed out earlier in this chapter, and as mentioned several times in this book. (Related reminder: please also notice that one could clearly perceive and explicitly realize, even merely from the large perspective or basic features of quantum mechanics itself, the basic *fact* that quantum mechanics is really unable to solve the most fundamental problem in front of itself—*why* there are quantum states, as clearly mentioned earlier in this chapter, and as explicitly pointed out several times in this book.) Specifically, this inability is explicitly reflected in the fact that quantum mechanics doesn't have the ability to reveal the *quantum* mechanism of why and how photons, being the tiny and discrete quanta or particles of light, get their velocity c (the speed of light) from the electron that emits them. Therefore, this kind of inspiration turns out unable to be truly valid, either in essence or in truth or in both. (Commentator: with this preparation, one shouldn't be too surprised even if it turns out that the so-called gravitons really don't exist.)

The second step is to answer the question: do the guessed gravitons really exist? The existence of gravitons requires a *prerequisite*. If this *prerequisite* is satisfied, one can say that gravitons may exist. On the other hand, if this *prerequisite* can't be satisfied, one has to say that gravitons don't exist. Since gravitons are the product of the assumed quantum theory of gravity (the other name for this theory is simply as quantum gravity theory or quantum gravity), clearly the *prerequisite* for the existence of the guessed gravitons lies with the existence of this assumed theory.

However, this assumed theory doesn't and can't exist in fact, simply because the precondition or the indispensable requirement for its existence can't exist at all, being clearly and sufficiently determined by the following hard evidence. Because the main purpose of the assumed quantum theory of gravity is to combine general relativity and quantum mechanics, clearly, even obviously, the precondition for the existence of this assumed theory is to solve the serious inconsistency of general relativity and quantum mechanics, so that the combination of them doesn't cause obviously ridiculous results, such as mathematical inconsistencies, meaningless infinite and negative probabilities. Nevertheless, as discussed and pointed out earlier, it is not possible to solve this serious inconsistency, simply because there is utterly no way to meet the necessary requirement for solving it. Thus, this assumed theory doesn't and can't exist in fact; that is, it turns out that the very *prerequisite* for the existence of the so-called gravitons doesn't exist at all. Therefore, we have no choice but to draw the clear and solid, actually unavoidable, conclusion: the guessed gravitons don't and can't exist at all. (Related question and answer: what is the key to grasping this conclusion easily and quickly? Answer: the so-called gravitons turn out to have been "invented" or fabricated from an actually non-existent theory, the assumed quantum theory of gravity.) (In addition, since the guessed gravitons don't exist at all, another assumption, which says that gravitons travel at the speed of light, turns out to be not only meaningless, but also obviously ridiculous!)

With this clear and solid conclusion—the guessed gravitons don't and can't exist at all, and simply because these ongoing theories (including string theory, along with its variants, as well as the supergravity theories) *indispensably* depend on the existence of the so-called gravitons, it thus becomes clear, even obvious, that these ongoing theories turn out unable to be correct or valid, either in fact or in essence. This is the irrefutable or undeniable *fact* in front of these ongoing theories; this is the inescapable *reality* that these ongoing theories have to face up to! (Narrator: after seeing such a "cruel" but undeniable fact, and after realizing such a "harsh" but inescapable reality, disappointment then hopelessness occupies the mind of the owner of this seemingly wonderful building. But it's not too bad; he soon begins to be deeply lost in his meditation and pondering.)

A few weeks later, this owner has made a calm and reasonable decision from his deep contemplation. One day, while he is walking along a quiet and secluded road with both sides full of dense trees, he keeps talking to himself: "Are there available theories that have solved the problems I am facing?" "If there are such theories, they ought to be able to tell us: why space and time are variable thus relative in a gravitational field, why there are quantum states; and they can also reveal the certainty mechanism behind the uncertainty principle." "I desperately need these kinds of new theories." He seems to forget time while walking and thinking, thinking and walking. Surprisingly, he runs into his contractor (the contractor mentioned at the start of this chapter)! The owner asks: "Oh, what are you doing here?" "I am thinking the previous methods and techniques might not be able to fix the severe problems of your building, so I am looking for new methods and techniques," the contractor answers honestly. This owner becomes so happy and excited that he invites: "Let's walk together and have a talk."

CHAPTER 10

NOT WEIRD IF DIFFICULT TO UNDERSTAND "A BRIEF HISTORY OF TIME"

[The Window on This Chapter]

Having been aware of the secrets of why it's difficult to understand the book *A Brief History of Time*, should people still feel disappointed or discouraged for having not really understood that book?

After the secrets of why it's difficult to understand *A Brief History of Time* have been revealed, should *A Brief History of Time* complain as if *it* has been treated unfairly, or is *it* supposed to have the feelings of inferiority or self-pity?

What are the simple and effective ways/methods to convince *A Brief History of Time* not to complain and not to have the feelings of inferiority or self-pity?

Why should *A Brief History of Time* be highly valued and fully respected from a large perspective or from a long-term point of view?

Are there great inspirations or profound implications from the reality that it's difficult to understand *A Brief History of Time*? What are they?

Ladies and gentlemen: the answers to these questions will be unfolding in this chapter!

The previous chapters have revealed the secrets of why it's extremely difficult to understand *A Brief History of Time*! After revealing these secrets, they become open secrets of course; and they are no longer secrets. After this revealing, it turns out that a series of insurmountable difficulties in understanding that book didn't arise from the people who had read it, because (it turns out that) these difficulties directly came from that book itself. This means that one shouldn't have felt discouraged for having not really understood *A Brief History of Time;* this indicates that one shouldn't have felt disappointed for having encountered many great challenges in trying to understand that book. In other words, knowing these secrets turns out to be a wonderful relaxation or great consolation for the people who have been perplexed or struggled for having not really understood *A Brief History of Time*. (Commentator: in fact, to those who have fully realized these secrets, it seems to be quite reasonable and fair that they can naturally, or easily, come to such a clear conclusion: perhaps, no one could really understand *A Brief History of Time*. That is, it turns out to be rather rational thus quite understandable that one couldn't really understand *A Brief History of Time*.)

Having revealed these secrets, it turned out that it's actually impossible to imagine that one could really understand *A Brief History of Time (ABHOT,* for short)—a clear *reality;* it turned out to be neither reasonable nor feasible to believe that one could truly understand *ABHOT*—a plain *fact*. In the face of such a clear *reality* and such a plain *fact*, *ABHOT* might have two possible reactions. One is to complain as if it has been unfairly treated or grievously misunderstood; another is full of the feelings of inferiority or self-pity. However, in this chapter I shall argue through providing sufficient reasons: *ABHOT* actually has nothing to complain; *ABHOT* ought not to have any feelings of inferiority or self-pity at all; instead *ABHOT* should be highly valued and fully respected when viewed from the large and far-sighted perspective of science development or progress. (While these arguments can and will make *ABHOT* feel happy and confident, the purpose of my arguments is not to flatter *ABHOT* at all, because these arguments, dear readers will see, are totally based on authentic, objective and rational grounds.)

How to dissipate the possible complaints from *A Brief History of Time?* How to convince that *A Brief History of Time* shouldn't complain? What effective and feasible measures should be taken to accomplish that? The task of exploring the answers to these questions, being one of the two focuses in this chapter, will be implemented step by step as follows.

For the people who have read *A Brief History of Time (ABHOT,* for short), they might have realized that the serious inconsistency of general relativity and quantum mechanics turned out to be a big obstacle to their really understanding *ABHOT*. (In fact, Hawking has clearly mentioned this inconsistency in his *ABHOT*—P. 12, 1998 version, which could be a noticeable psychological preparation for people to face the serious inconsistency of general relativity and quantum mechanics. Such a preparation is actually a considerable help for people to think over and recognize this serious inconsistency.) Moreover, the related analyses and discussions in previous chapters have also revealed and shown: this serious inconsistency, along with its profound or actual implications, is indeed the impassable barrier to anybody who wants to have a real understanding of *ABHOT*. (Narrator: that is, this serious inconsistency, together with its unavoidable implications, turns out to be the root cause or "prime culprit" that practically rules out the possibility that one could really understand *ABHOT!*) Thus, as long as *ABHOT* can completely understand this serious inconsistency—this will also enable *ABHOT* to be well aware of the corresponding implications of this serious inconsistency, all its possible complaints, caused by the *reality* that it turns out to be reasonable and fair that one couldn't really understand *ABHOT,* will soon dissipate. And so, convincing *ABHOT* to digest and acknowledge this serious inconsistency turns out to be an ideal approach of dispersing all its possible complaints. Therefore, seeking the answers to the questions above actually turns into looking for the answer to such a question: what are the simple, concrete and effective ways or methods of plainly and easily understanding the serious inconsistency of general relativity and quantum mechanics, from new viewing angles? (Or how could *ABHOT* clearly sense thus explicitly realize this serious inconsistency without reluctance?)

In order to find out the answer to this question, one needs to be familiar with the basic attributes of both general relativity and quantum mechanics. General relativity describes the phenomena on large-scale objects such as stars and galaxies, as well as the universe, whereas quantum mechanics deals with the phenomena on extremely small scales about the objects of tiny things, like atoms and electrons. So these two theories involve two *different* types of phenomena, or *different* phenomena, to be simpler (that is, general relativity and quantum mechanics are two separate theories working on the very different phenomena or things in quite different situations). This feature can direct us to explore effective and simple measures that will substantially help us in our clearly and easily understanding the serious inconsistency between general relativity and quantum mechanics, through the following two plain and straightforward questions.

First question: are the two aspects of the same phenomenon, such as light—a fundamentally important phenomenon in science, factually compatible thus actually consistent with each other (within the paradigm of quantum mechanics along with the uncertainty principle from it)? If the answer is NO to this question, then one shouldn't be surprised towards the serious inconsistency of quantum mechanics and general relativity—people can easily face this serious inconsistency. Why? How? Since even merely the two aspects of the *same* phenomenon turn out to be factually incompatible thus actually inconsistent with each other, needless to say two separate theories that involve *different* phenomena. This is a bit like: if the two hands of one person cannot cooperate or coordinate well—being analogous to that the two aspects of the *same* phenomenon are actually inconsistent with each other, then it is quite natural that the four hands of two people cannot cooperate or coordinate well—being analogous to that two separate theories working on *different* phenomena are seriously inconsistent with each other. (Commentator: yes, that is clearly valid and definitely true. This is because it is clear, even obvious, that the requirement for the consistency of the two aspects of the *same* phenomenon is much easier to meet than for the consistency of two separate theories dealing with *different* phenomena. That is to say, compared to that the two aspects of the *same* phenomenon are actually inconsistent with each other, the serious inconsistency of two separate theories that work on *different* phenomena is or becomes pretty reasonable and fair to most people.)

Second question: are the two core concepts from the same theory (special relativity, for example) factually compatible thus obviously consistent with each other? If the answer is NO to this question, then the serious inconsistency of general relativity and quantum mechanics is nothing to be surprised about—one could easily face this serious inconsistency calmly and rationally. Why? Since even just the two concepts of the same theory that deals with the objects or things in the *same* situation turn out to be factually incompatible thus obviously inconsistent with each other, needless to say two separate theories that involve the objects or things in quite *different* situations. (Narrator: clearly, the requirement or constraint on the consistency of the two concepts from the same theory for the *same* situation is not as strict as on the consistency of two separate theories for *different* situations. This is a bit like: if the consistency of the former is analogous to that one person walks in steady and constant steps, then the consistency of the latter can be quite comparable to that two people simultaneously walk in steady, constant, and consistent steps. Thus, in comparison with that the two concepts from the *same* theory for the *same* situation are obviously inconsistent with each other, the serious inconsistency of two separate theories for *different* situations is or appears rather rational, thus quite natural or reasonable in the eyes of most people.)

Quite obviously, also rather rationally, the above two questions, together with the analyses and discussions to them, unfold two simple, effective and feasible measures that can explicitly and considerably help one perceive and grasp the serious inconsistency of general relativity and quantum mechanics clearly, tangibly and easily. So these measures will be the principles or criteria that guide our following discussions. (Commentator: thus, the first consolation is coming towards *A Brief History of Time!*)

As the first action to carry out these measures, we shall work on such a subject: why and how the well recognized concept of the wave/particle duality of light can specifically help us understand the serious inconsistency of quantum mechanics and general relativity. This concept says that a ray of light has both wave and particle natures but not at the same time—a ray of light can be thought as a wave or as a particle, but cannot be thought as both at once. So this concept has the characteristic feature that the choice of one description precludes the simultaneous choice of the other. As a preparation for this action, we need to analyze and discuss the essence and actual implication of this concept.

What is the *essence* of the concept of the wave/particle duality? The essence is that the wave-particle dual natures of light cannot be unified, or at least cannot be within the paradigm of quantum mechanics (along with the uncertainty principle based on it). This is clear and self-evident! If the dual natures of light could be unified, they would have already been unified; if the dual natures of light could be unified, who would have looked for trouble by developing the complicated concept or theory of the wave/particle duality?! Then, why is it unable to unify the wave-particle dual natures of light? This question is more or less equivalent to ask: what is the actual or deepest implication of the concept of the wave/particle duality?

The answer is: the wave nature and particle nature of light are factually incompatible thus actually inconsistent with each other. This answer can be clearly seen through the following three hard facts or undeniable realities. First, the particle nature of light doesn't require any identifiable physical medium for its propagation, such as the light from the sun to the earth. On the contrary, the wave nature of light must require a medium for its propagation, like sound waves going through air. To be clarified, the wave nature of light *on its own* is utterly unable to explain the fact that light can travel through totally empty space—this kind of space doesn't have recognizable physical medium for propagating light waves. And the famous historical event about looking for the "ether" that happened in the second half of the 19th century might help remind us of this point; the claimed "ether" was then believed as the medium propagating light waves. (Narrator: according to its wave nature, light couldn't travel through totally empty space, whereas according to its particle nature, light could. Then if one asks light: "Can you travel through totally empty space?" "I don't know," or "I can't know," is probably, and should be, the rational or sensible response from light. Clearly, such a kind of response is actually a confirmation of the answer above. Certainly, a ray of light from the sun to the earth cannot tell us: "Only my particle nature comes, but my wave nature still remains on the sun.") Second, the wave nature and particle nature of light are used *separately* in explaining different phenomena. Specifically, the interference and diffraction of light can only be interpreted with the *wave* nature of light, whereas only the *particle* nature of light can explain the energy radiation from a hot object or body (like an oven or a star), the photoelectric effect (the phenomenon of ejecting electrons from a metal shone by high-energy light), and the Compton scattering (the experiment in which the high-energy rays of light, X-rays, were used to strike electrons). Third, the wave nature and particle nature of light are actually not compatible, because particles are localized, waves are spread out; thus conceptually one cannot blend them together. In fact, it is not possible to derive the quantum theory of light from the wave theory of light, or vice versa; that is, there seems to be an impassable wall between the wave/particle dual natures of light. All in all, the above realities clearly and consistently point to or demonstrate such a concrete *fact:* indeed, the wave nature and particle nature of light turn out to be factually incompatible thus actually inconsistent with each other.

This concrete fact can explicitly and considerably broaden our field of vision on viewing the serious inconsistency of quantum mechanics and general relativity. Why? How? With this concrete fact as a clear reminder, one can't help thinking of or thinking over: since only just the two aspects of the *same* thing (a ray of light in this case) turn out to be factually incompatible thus actually inconsistent with each other within the paradigm of quantum mechanics (along with the uncertainty principle based on it), it seems neither rational nor realistic to ask for the consistency of quantum mechanics and general relativity, because they cover *different* phenomena, or *different* things, as just mentioned above; the serious inconsistency of these two separate theories thus appears to be rather rational, at least not irrational at all! This is a bit like: if the hands and feet of one person can't cooperate or coordinate well—being analogous to that the two aspects of the same thing are actually inconsistent with each other, then it is quite normal that the hands and feet of two people can't cooperate or coordinate well—being analogous to that two separate theories working on *different* things are seriously inconsistent with each other. As a result, such an explicit and considerable enlargement of our field of vision is really a noticeable and substantial improvement in our viewing and thinking about the serious inconsistency of quantum mechanics and general relativity. (Commentator: yes, that is clearly true. This is because the requirement for the consistency of two separate theories that deal with *different* things is much more stringent than for the consistency of the two aspects of the *same* thing. Thus, compared to that the two aspects of the *same* thing are actually inconsistent with each other, the serious inconsistency of two separate theories that work on *different* things is or becomes pretty reasonable and fair to most people.)

The existence of this concrete fact is also an explicit and substantial help for *A Brief History of Time (ABHOT)* to dissipate its possible complaints. Why? With the clear and noticeable reminder from this concrete fact, *ABHOT* can easily perceive that the serious inconsistency of quantum mechanics and general relativity is not only a well-known fact, but also a rather rational or quite natural reality. With this perception, *ABHOT* can clearly realize that the existence of this serious inconsistency turns out to be reasonable and fair. This clear realization can effectively remove or substantially reduce the psychological hesitation or reluctance of *ABHOT* when facing this serious inconsistency, which is very helpful for *ABHOT* to have a good understanding of this serious inconsistency. Please be reminded that, as discussed and pointed out earlier in this chapter, the key for *ABHOT* to dissipate its possible complaints lies with that it can completely understand this serious inconsistency. (Narrator: thus from this moment, the first consolation begins to take effect in the mind of *ABHOT*! From now on, all the possible complaints of *ABHOT* will be gradually disappearing, like air leaking out of a balloon.)

Apart from the targeted purpose of this chapter, the concrete *fact*, which is that the wave nature and particle nature of light turn out to be factually incompatible thus actually inconsistent with each other, can urge or remind us to inspect some related concepts in science and rethink the mode of thinking incubated from these concepts, especially when this *fact* is viewed from a large perspective. For instance, it seems pretty reasonable and fair that this *fact* can naturally lead one to think of or ask the following related question. Since even just the two aspects of the *same* phenomenon, light, are factually incompatible thus actually inconsistent with each other, thereby cannot be unified, could it be realistic to engage in the theme of "the unification of physics" that necessitates unifying *different* physical theories that cover a wide range of *different* phenomena? One may easily think over this question and get the answer to it via the following simple analogy as a concrete comparison or reminder.

Let us say, theory A describes one phenomenon that has two aspects, and theory B deals with another phenomenon that also has two aspects. Clearly, even obviously, in order to unify these two theories, one must first unify the two aspects of each of both the phenomena (without such a unification, it is definitely not possible to unify these two theories). By this plain reasoning, it seems that one shouldn't have much difficulty reaching such a rational conclusion: the dream of "the unification of physics" cannot be true in reality, or at least cannot be true within the paradigm of modern physics with quantum mechanics and general relativity as its main theoretical pillars. (One can easily grasp this conclusion or chew it over via the following simple and clear situation as a vivid analogy. An experienced tailor must have known such a basic principle: because an entire piece of clothing is made up of several different parts, if the inside and outside layers of each part were inconsistent with each other, it would be quite natural that the entire piece of clothing cannot be successfully integrated or satisfactorily connected when these different parts are put together; it would be neither rational nor realistic to expect that the entire piece of clothing can be successfully integrated or smoothly connected when these different parts are pieced together.) (Commentator: the most dangerous enemy of science is wrong concept and incorrect thinking mode! Science has nothing to fear but fear wrong concept and incorrect thinking mode! This has been clearly shown and impartially witnessed by the tortuous history of science development; this tortuous history is, of course, very familiar to all today's intelligent or knowledgeable scientists. In fact, it seems to be quite reasonable and safe if one concludes that most educated people, even having just experienced middle or high schools, are also somewhat familiar with this tortuous history.)

In addition to what we have seen above, the analyses and discussions above may also bring us two closely related questions. The essence and the real implication of the concept of the wave/particle duality of light actually tell us such a basic fact: because the wave/particle dual natures of light are factually incompatible thus actually inconsistent with each other, they cannot be unified at all within the paradigm or stereotype of quantum mechanics along with the uncertainty principle based on it. This basic fact can easily lead some insightful and attentive readers to think of or raise some fundamentally important questions like: can we eventually tackle the problem that the wave/particle dual natures of light are factually incompatible thus actually inconsistent with each other? And after solving this problem, can we ultimately unify the dual natures of light? The answer is YES to both of these two questions. (Commentator: if your answer was NO or I DON'T KNOW, then some people could question the validity of your arguments above.)

But this YES can be materialized only through a radically new theory that is able to find out and reveal the *mechanism* behind the phenomenon of the wave/particle duality of light (this new theory has been concisely mentioned in chapter six as a newly discovered physical law). This new theory shows that: after revealing the common *mechanism* behind the two different aspects (wave aspect and particle aspect) of light, these two aspects turn out to be essentially consistent thus naturally unified, thereby realizing the historic unification of the wave/particle dual natures of light. (Commentator: so this historic unification is also a clear indication that in science, it is fundamentally important or crucially necessary to reveal the mechanism behind an observed phenomenon, instead of merely staying on the surface of describing and interpreting the observed phenomenon. In fact, one of the most important tasks in science is to reveal the mechanism thus grasp the essence behind observed phenomena, being clearly and objectively determined by the fundamental nature and ultimate goal of science.) (For the detailed information about the profound implications and fundamental significance of this historic unification, please go back to the related sections or parts of chapter six if necessary.)

What comes along with this profound and historic unification is a clear and noteworthy clue that directly points to such a candid reality: what quantum mechanics cannot solve doesn't necessarily mean other theories cannot either! With this candid reality as a clear and constant reminder, it seems not difficult for *A Brief History of Time (ABHOT)* to be aware that quantum mechanics is, at least, not a complete theory, which can naturally, or easily, lead *ABHOT* to realize: quantum mechanics may be a theory that has severe flaws, so its serious inconsistency with general relativity ought not to be a big surprise at all. Undoubtedly, such a realization is substantially conducive for *ABHOT* to face and digest the serious inconsistency of quantum mechanics and general relativity, which is a noticeable help for *ABHOT* to dispel its possible complaints. (Commentator: thus, after noticing this clear clue, *ABHOT* also feels further relieved, at least in a sense or from a certain angle.)

Perhaps what we have done above is still not enough for *A Brief History of Time* to dissipate all its possible complaints; it is anxiously waiting for other action to disperse its remaining complaints.

Well, as the second action to implement the simple, effective and feasible measures mentioned earlier, we shall discuss such a topic: why and how the *reality*, which is that the two core or key concepts of the same theory turn out to be factually incompatible thus essentially contradictory, therefore being obviously seriously inconsistent with each other, is actually a good window on viewing the serious inconsistency of general relativity and quantum mechanics. Through such a window, one can see this serious inconsistency more clearly, perceive it more explicitly, and realize it more definitely. In this coming action, Einstein's theories of special relativity and general relativity will be analyzed and discussed as concrete examples.

Special relativity has two core or key concepts, which are length contraction and time dilation (respectively for interpreting length becomes shorter and time runs slower that appear at high speed). The detailed analysis and discussion in chapter two have uncovered or shown such an unavoidable *reality:* length contraction and time dilation turn out to be factually incompatible thus essentially contradictory with each other, because they would directly deny each other if met together; that is, length contraction and time dilation turn out to be obviously seriously inconsistent with each other. This *reality* is, of course, a plain manifestation of such an inescapable *fact:* the two concepts of the same theory for the same situation turn out to be obviously seriously inconsistent with each other. With the considerable and explicit help from this *reality* and this *fact*, A Brief History of Time (*ABHOT*) can easily perceive and realize the serious inconsistency of general relativity and quantum mechanics more clearly and definitely, via the simple and related reasoning or thinking as follows. Since even merely the two concepts from the same theory (that deals with the objects in the *same* situation) turn out to be obviously seriously inconsistent with each other, the serious inconsistency of two separate theories that involve the objects in *different* situations is or becomes rather rational, thus pretty reasonable and fair (please be reminded that general relativity and quantum mechanics are two separate theories dealing with the phenomena or objects in quite different situations, as mentioned earlier). As a result—as a clear and pleasurable result in effect, this unavoidable *reality* and this inescapable *fact*, because they can make or help *ABHOT* distinctly sense and explicitly realize the serious inconsistency of general relativity and quantum mechanics without reluctance, are certainly the effective ways or channels for *ABHOT* to dispel its remaining complaints. (Narrator: clearly, even obviously, the requirement or constraint on the consistency of two separate theories for *different* situations is much more difficult to meet than on the consistency of the two concepts from the same theory for the *same* situation. Therefore, in comparison with that the two concepts of the same theory for the *same* situation are obviously seriously inconsistent with each other, the serious inconsistency of two separate theories for *different* situations is or appears rather rational, thus quite reasonable in the eyes of *ABHOT*.)

More directly thus probably more effectively, the above-mentioned unavoidable *reality* and inescapable *fact* in front of special relativity can drive away the remaining complaints of *A Brief History of Time* more quickly, through general relativity as an outlet. Why? The reason is simple and clear: general relativity is crucially based on special relativity. Specifically, in the face of the *reality* and *fact* mentioned above, and since general relativity is indispensably dependent on the length contraction and time dilation of special relativity, believe it or not, general relativity actually has no other options but the following two "choices".

One choice is to accept the length contraction in a gravitational field, for being seemingly consistent with the concept of length contraction that comes from special relativity. (General relativity has already agreed that time runs slower in a gravitational field, which is the product of its having admitted the time dilation in such a situation.) This acceptance, however, results in the clear and serious inconsistency of the length contraction and the time dilation of general relativity, like the two core concepts of special relativity just mentioned above. In other words, this acceptance is actually the option that causes the two crucial concepts of general relativity to be fatally inconsistent with each other. Therefore, this choice turns out to be a clear signal that directs *A Brief History of Time* to face and digest the serious inconsistency of general relativity and quantum mechanics: since only just the two crucial concepts from the *same* theory turn out to be fatally or seriously inconsistent with each other, the serious inconsistency of general relativity and quantum mechanics is nothing to be surprised about at all. Clearly, even obviously, unless the two crucial concepts of a theory are consistent with each other, it is impossible or extremely difficult to imagine the theory can be truly consistent with other theories, or it is rather rational that the theory is inconsistent with other theories. (This is a bit like that an experienced engineer must have been familiar with such a basic or plain principle, also a self-evident rationale: if the two sides of a wall are seriously inconsistent with each other, it is very natural or quite normal that this wall is inherently and permanently inconsistent with others.) (Narrator: undoubtedly, this choice is an effective outlet for *A Brief History of Time* to dispel its remaining complaints.)

Another choice of general relativity is to reject the length contraction in a gravitational field, in order to avoid the aftermath of the choice above. This rejection, however, is equivalent to refusing the length contraction of special relativity (because it is the length contraction embedded in special relativity that is ultimately responsible for interpreting the length contraction in a gravitational field, if general relativity doesn't reject the length contraction in such a situation). Therefore, this rejection is the direct result of general relativity has perceived the unavoidable outcome that general relativity itself is incompatible with special relativity. That is, this choice is, in fact, equivalent to admitting that general relativity is inconsistent with special relativity. As a result, this choice turns out to be also a noticeable sign that guides *A Brief History of Time* to face and digest the serious inconsistency of general relativity and quantum mechanics: since even the two theories on the *same* type of phenomena (both special relativity and general relativity deal with the concept of space and time, though in different situations) turn out to be actually inconsistent with each other, the serious inconsistency of two separate theories that involve *different* types of phenomena appears rather rational or normal, at least not irrational at all—general relativity and quantum mechanics are two separate theories dealing with two different types of phenomena, as mentioned earlier. (Narrator: clearly, this choice is another effective outlet for *A Brief History of Time* to disperse its remained complaints.)

After reading the information in the several paragraphs above, some readers might have realized: your arguments are valid, but you can make these arguments even more convincing, through a sharp contrast or comparison, if you can answer the following two closely related questions.

First question: is there a new theory that has not only explained *why* length becomes shorter and *why* time runs slower at high speed, but also shown that these explanations are completely compatible thus consistent? The answer is YES. This new theory is the newly developed and verified, mechanism-revealed scales relativity theory that has two crucially important concepts: length scale reduction and time scale reduction that appear in the situation of high speed (about these two crucial concepts, please go back to chapter two if necessary). These two concepts, by revealing and determining why and how the scales of length and time are reduced at high speed, clearly tell us *why* length becomes shorter and *why* time runs slower in the situation of high speed. And more importantly, these two concepts, as the result of having answered these two *whys* by revealing the mechanism behind them, are fully compatible thus completely consistent with each other! (Related questions and answer: can this question and the answer to it give a specific help for *A Brief History of Time (ABHOT)* to face and understand the serious inconsistency of general relativity and quantum mechanics? Why? Answer: yes, they can. With this question as well as the answer to it, and through a simple comparison, it seems not difficult that *ABHOT* can clearly see or sense such a naked truth: indeed, special relativity is unable to solve the fundamental problem in front of itself, *why* length becomes shorter and *why* time runs slower at high speed. With and through this naked truth, *ABHOT* can evidently and surely realize: special relativity turns out to be a theory that inherently has severe flaws. With such a definite realization kept in mind, along with the constant reminder that general relativity is crucially based on special relativity, *ABHOT* can easily and clearly perceive: the serious inconsistency of general relativity and quantum mechanics turns out to be quite natural or reasonable, thus rather understandable. Such a perception is certainly a considerable help for *ABHOT* to have a clearer and better understanding of this serious inconsistency. As a result, this question and the answer to it are well able to provide a specific help for *ABHOT* to face and understand this serious inconsistency, therefore making *ABHOT* speed up dispelling its remaining or possible complaints.)

Second question: is there some new theory that, with its new concepts, has not only revealed *why* length becomes shorter and *why* time runs slower in a gravitational field (to the observers far away from it), but also shown that these new concepts are totally compatible thus consistent? The answer is also YES. Such a new theory is the newly discovered and verified, mechanism-revealed gravitational theory that has two core concepts: gravitational length scale reduction and gravitational time scale reduction that take place or appear in a gravitational field (about these two core concepts, please go back to chapter three if necessary); the innermost essence of this new theory is that the existence of a gravitational field causes a certain reduction in the scales of space and time in the gravitational field, along with that: the stronger the gravitational field, the larger the reduction. These two concepts, by revealing and determining why and how the scales of space and time are reduced in a gravitational field, evidently show *why* length becomes shorter and *why* time runs slower with respect to the observers far away from the gravitational field. Moreover, these two core concepts, as a result of having answered these two *whys* by revealing the mechanism behind them, are inherently compatible thus fully consistent with each other! (Related questions and answer: can this question and the answer to it give a substantial help for *A Brief History of Time (ABHOT)* to digest and grasp the serious inconsistency of general relativity and quantum mechanics? Why? Answer: yes, they can. Having witnessed this question as well as the answer to it, it seems rather rational that *ABHOT* can distinctly and explicitly perceive, through a sharp contrast, such a clear fact: general relativity turns out to be a theory with inherent and severe flaws, simply because it doesn't have the ability to solve the very basic problem it has to face, *why* length becomes shorter and *why* time runs slower in a gravitational field. Such a clear and definite perception can substantially help *ABHOT* further realize that the serious inconsistency of general relativity and quantum mechanics turns out to be rather reasonable and fair, thus quite understandable. Clearly, even obviously, this realization is certainly a considerable and noticeable help for *ABHOT* to face this serious inconsistency with a positive and rational attitude. Therefore, this question and the answer to it are definitely a substantial help for *ABHOT* to digest and grasp this serious inconsistency, thereby causing *ABHOT* to disperse its remaining or possible complaints more quickly.)

Up to here, and up to now, in order to dissipate all the possible complaints from *A Brief History of Time (ABHOT)* caused by the *reality* that it turns out to be both reasonable and realistic that one couldn't really understand *ABHOT*, we have taken various actions and gone the extra mile along the route of convincing *ABHOT* to face, digest and comprehend the serious inconsistency of general relativity and quantum mechanics. Through what we have done above, it seems not difficult that *ABHOT* has comprehensively and thoroughly understood, from different angles, depths and heights, this serious inconsistency, thereby clearly and fully realized its profound and extensive implications. As analyzed and pointed out at the beginning of this chapter, as long as *ABHOT* can completely understand this serious inconsistency, thus explicitly know its corresponding implications, all the possible complaints of *ABHOT* caused by this *reality* will soon dissipate. (Commentator: so at this point, most likely all the possible complaints of *ABHOT* have dissipated, because there are several channels or outlets through which these complaints have been dispersed very quickly, a bit like air leaking out of a balloon through several holes.)

After *A Brief History of Time* has emptied itself by dissipating all its possible complaints (through fully understanding the serious inconsistency of general relativity and quantum mechanics, thus being well aware of the corresponding implications of this serious inconsistency), with what ought it to refill itself again? Is it a feeling of inferiority? NO! Is it a reaction of embarrassment? NO! Is it a sentiment of self-pity? NO! After convincing *A Brief History of Time* to dispel all its possible complaints, with what ought we to refill it? And how should we refill it? Should we refill it with a low opinion of it? NO! Should we refill it with an underestimate of it? NO! Should we take pleasure in its misfortune? NO! (Related question and answer: it seems that these negative answers in themselves, that is, the *six* NOES in a row, couldn't be satisfied for most earnest or curious readers, do you have specific and adequate reasons to convince them? Answer: yes, I have. I shall provide *seven* good reasons from the large and long-term perspective of the great roles that *A Brief History of Time* has actually played, even though these roles may not be its intended or original purpose at first sight.)

(Reason 1) most of all, or overall, *ABHOT* (the initials of *A Brief History of Time*) shouldn't look down upon itself because of the "harsh" *reality* that it turns out to be both reasonable and fair that one couldn't really understand *ABHOT*. Why? It was *ABHOT* that, just through such a *reality*, has impartially witnessed the serious inconsistency of general relativity and quantum mechanics. And it is just this serious inconsistency that has led us to find out the direct reason why these two theories are permanently inconsistent, and that has guided us to dig out the deepest reason why these two theories are inherently inconsistent (for the specific information about these two reasons, please go back to chapter one if necessary). After finding out this direct reason and after digging out this deepest reason (in chapter one), it turns out that this serious inconsistency is not only unavoidable but also irreparable. As a result, believe it or not, this serious and irreparable inconsistency shakes the large mansion of modern physics in effect, or at least shows that this mansion cannot be stable in fact, because general relativity and quantum mechanics are the two main theoretical pillars of modern physics. Thus, *ABHOT,* through this "harsh" *reality*, may play a positive role in reminding us to realize or notice the possible clues that can point to a great and historic revolution in physics. (Narrator: for this great and positive role, we shouldn't have a low opinion of *ABHOT*, even though it has to face the *reality* that it turns out to be both rational and fair that one couldn't really understand *ABHOT*.)

(Reason 2) *A Brief History of Time (ABHOT)* shouldn't feel embarrassed by the serious and irreparable inconsistency of general relativity and quantum mechanics, even though such an inconsistency turns out to be the root cause or "prime culprit" of the *reality* that it turns out to be both reasonable and realistic that one couldn't really understand *ABHOT!* Why? It was just through this very inconsistency that *ABHOT* has objectively witnessed such a solid *fact:* either general relativity or quantum mechanics or both cannot be correct! (Friendly reminder and/or impartial commentator: if some dear readers, especially some of those respected professional people in physics, are still psychologically hesitant or reluctant towards this solid fact, please notice such an explicit fact that Dr. Hawking has clearly pointed out in his *ABHOT:* general relativity and quantum mechanics cannot both be correct because they are known to be inconsistent with each other—1998 version, P. 12. And please also notice that this solid fact is totally consistent with this explicit fact. Therefore, the clear existence of this explicit fact can effectively remove one's psychological hesitation or reluctance towards this solid fact, which can enable him or her to face this solid fact rationally and objectively. In fact, it is both rational and safe if one concludes that all excellent physical experts are very familiar with this explicit fact, because they know the serious inconsistency of general relativity and quantum mechanics quite well; because this explicit fact directly comes from and is clearly determined by this serious inconsistency. Moreover, this serious inconsistency is so well known that it has become very important general knowledge or common sense in physics; accordingly, this explicit fact, being the obvious and unavoidable outcome or product of this serious inconsistency, has also become a well-known or fully recognized fact—at least to the professional people in physics. All in all, with the substantial and noticeable help from this explicit fact, it seems to be both reasonable and fair if one comes to the conclusion that most people, even if not everyone, can face this solid fact calmly and rationally.)

With this solid *fact* as a clear clue, and after digging out the deepest reason behind this inconsistency (in chapter one), it turns out that both these theories, general relativity and quantum mechanics, cannot be correct, because neither of them is able to solve the most fundamental problem in front of itself, as explicitly pointed out and/or clearly mentioned several times in this book. Consequently, believe it or not (admit it or not), this serious, unavoidable and irreparable inconsistency actually shakes the entire mansion of modern physics—drastically and inevitably, given that these two theories are the two main theoretical pillars of modern physics.

Thus, one can see that *A Brief History of Time*, through such a kind of inconsistency, (though perhaps by accident) can play a pioneering role in hinting us to become aware of the signs that may forecast the coming of a great revolution in physics. (Narrator: for such a possible role, it seems that we ought not to underestimate *A Brief History of Time* from a long-term and large perspective. And more importantly, this possible role, when considered together with the similar inspiration coming from the two well-known, also quite comparable, historical facts with profound implications and great significance, can become further realistic. What are these two facts then? Let us concisely review and discuss them as follows.)

Some people, especially those professional people in physics, may still remember the two famous clouds over the head of classical physics in the second half of the 19th century. One cloud was the fact that Newton's classical idea of absolute time and space was unable to interpret the observed phenomenon that the speed of light is the same for all observers, regardless of their motion relative to the source of light. This fact shook the foundation of classical physics, thereby led to the birth of Einstein's special relativity in 1905. Another cloud was the fact that the classical theory of electromagnetic waves was unable to explain the energy radiation from a hot object or body, like an oven or a star. This fact also shook the foundation of classical physics, thus brought about the birth of Planck's quantum theory in 1900. Not only that, both today's experts in physics and the historians in science have such a broad consensus: those two clouds or "old" facts had played a decisive or leading role in the great revolution from classical physics to modern physics in the early 20th century.

Today, after digging out the deepest reason behind the serious and irreparable inconsistency of general relativity and quantum mechanics (the specific information about this deepest reason has been presented in chapter one), there are two facts in front of us again. One is that general relativity is unable to resolve the most fundamental problem in front of itself, *why* space and time are variable thus relative in a gravitational field, simply because it is incapable of revealing the mechanism behind this *why* (that is, general relativity doesn't and can't have the ability to reveal the mechanism of *why* length becomes shorter and *why* time runs slower in a gravitational field). Another is that quantum mechanics can't resolve the most fundamental problem in front of itself either, *why* there are quantum states, simply because it is unable to reveal the *quantum* mechanism of why and how photons, being the tiny and discrete quanta or particles of light, get their velocity c (the speed of light) from the electron that emits them. Moreover, in comparison with those two "old" facts, these two "young" facts are obviously or actually much more fundamentally and crucially important, because general relativity and quantum mechanics, being the two main theoretical pillars of modern physics, are fundamentally and crucially important to the fate and future of modern physics. (Commentator: since the two "old" facts had played a leading role in the great revolution from classical physics to modern physics, why can't today's two "young" and far more important facts play a pioneering role in the even greater revolution from modern physics to mechanism-revealed physics?) (After hearing the far-sighted, rational and objective, also genuine or sincere, positive comment like the above, *A Brief History of Time* not only throws away its embarrassment, but also brings back its calm and serenity.)

(Reason 3) *A Brief History of Time (ABHOT)* shouldn't have the feelings of inferiority, even though its most/many parts are based on general relativity that turns out unable to be correct. Why? It was *ABHOT* that, through the *reality* that it turns out to be reasonable and fair that one couldn't really understand *ABHOT*, has impartially confirmed the uncovered fact, which is the solid conclusion that general relativity turns out unable to be correct. (As analyzed and shown in chapter three, this solid conclusion is sufficiently demonstrated and determined by the hard evidence that general relativity is unable to solve the most fundamental problem in front of itself—*why* space and time are variable thus relative in a gravitational field, simply because it doesn't and can't have the ability to reveal the mechanism behind this *why* at all. For the other reasons leading to this solid conclusion, and for the pieces of related information on how to see this solid conclusion more clearly, please see the related sections or parts of chapter three.)

This impartial confirmation has at least two obviously important roles. One is to reduce the psychological reluctance of some people (especially some of those respected professional people in physics) in facing the uncovered fact that general relativity turns out unable to be correct in truth. Another is to urge or remind us to digest and accept the new theory that has solved the fundamentally important problem of *why* space and time are variable thus relative in a gravitational field (this new theory has been concisely mentioned in the later part of chapter three), since general relativity doesn't have the ability to solve this fundamentally important problem at all. Both these roles are the sonorous prelude for the upcoming, groundbreaking developments of physics and its closely related other sciences— such as astronomy, astrophysics (a branch of astronomy dealing with the physical and chemical structure of the stars, planets, etc.) and cosmology (the scientific study of the universe). (If some readers are skeptical of this prediction, then the radical developments and remarkable achievements in physics and its related other sciences in the 21st century can and will convince you. If one feels it's hard to imagine this prediction, please look back on the history of science development: this history has clearly shown us many amazing similarities by lots of rather similar, quite comparable and highly valuable experiences or historical instances, which can be used for reference.) With these remarkably important roles that can have a profound and far-reaching influence on the advancement of science, and that may have great and historic significance for the progress of science, *A Brief History of Time (ABHOT)* shouldn't feel inferior at all; and we ought to have a high opinion of *ABHOT* from a far-sighted perspective. (Narrator: yet what comes along with this confirmation is actually more than merely dispelling the feelings of inferiority of *ABHOT*. Why? The answer is coming.)

(Reason 4) the impartial and important confirmation above can remind, even urge, us to inspect the theory of special relativity. After being aware of the confirmation that general relativity can't be correct, it seems neither unusual nor unreasonable that one can't help thinking of or thinking about: since special relativity is the crucial foundation of general relativity, what is the fate of special relativity? Is the uncovered fact that general relativity turns out unable to be correct caused by or related to special relativity? With these questions as a specific clue or hint, and through the detailed analysis and discussion in chapter two, we have drawn such a clear and solid conclusion: special relativity can't be correct in fact, because it turns out to be clearly and seriously self-contradictory. (As revealed or shown in chapter two, the two core or key concepts of special relativity, length contraction and time dilation—respectively for interpreting length becomes shorter and time runs slower that appear at high speed, turn out to be factually incompatible thus essentially contradictory, because they would directly deny each other if met together. Consequently, also unavoidably, special relativity turns out to be clearly and seriously self-contradictory indeed.) As a result, *A Brief History of Time (ABHOT)*, via its confirmation that general relativity can't be correct through the "harsh" *reality* that it turns out to be reasonable and fair that one couldn't really understand *ABHOT*, can play a positive role in helping us face the newly revealed or shown fact that special relativity can't be correct either, at least in a sense or from the angles of this confirmation and this "harsh" *reality*. (Commentator: such a positive role, due to being substantially important and considerably helpful to the revolutionary progress or radical advancement of science with profound implications and great significance, should be highly appreciated from a long-term viewpoint or from the large perspective of science development.)

(Reason 5) *ABHOT*, standing for *A Brief History of Time*, shouldn't be ashamed because its many parts are based on or closely related to quantum mechanics that turns out unable to be correct. Why? It was *ABHOT* that, through the "harsh" *reality* that it turns out to be rational and fair that one couldn't really understand *ABHOT*, and through the serious and irreparable inconsistency of quantum mechanics and general relativity, may help us (especially some of those respected professional people in physics) have a reasonable attitude towards the new theories that have solved the crucially important problems quantum mechanics can't. For example, since quantum mechanics doesn't and can't resolve the fundamental problem of *why* there are quantum states, it seems not easy to find out an appropriate reason for not actively digesting and accepting the new theory that has solved *this* fundamental problem by revealing the *quantum* mechanism of *why* and how photons, being the tiny and discrete quanta or particles of light, get their velocity c (the speed of light) from the electron that emits them. Another example is that, since quantum mechanics is unable to reveal the *certainty* mechanism behind the *uncertainty* principle, it seems very difficult to have a valid excuse not to accept the new theory that has revealed and determined the *certainty* mechanism behind the *uncertainty* principle. Therefore, *ABHOT* can play a helpful role in reminding or guiding us, at least in a sense or to a certain degree, to digest and accept the related new theories for the sake of science. (Narrator: quite obviously, this possible role of *ABHOT* should be fully respected from the large and far-sighted perspective of science development and progress.)

(Reason 6) *A Brief History of Time*, through its witness to the serious and irreparable inconsistency of general relativity and quantum mechanics, can play an actively driving role in guiding or reminding us to accept the new theories that can solve the two famous, also pressing, big problems in science: *dark matter* and *dark energy*. (Commentator: some, even many, insightful and brilliant physicists begin to realize that, to a great extent, the fate of modern physics eventually hangs on whether or not it really has the ability to solve the big problems of *dark matter* and *dark energy*. Why? Most related experts have known, from the present observations, that *more than 95 percent* of all the matter and energy in the universe is ascribed to *dark matter* and *dark energy*! So it appears rather rational that modern physics has to face the great challenges from *dark matter* and *dark energy*. However, the unavoidable and inescapable *reality* is that *dark matter* and *dark energy* have actually become long-term unsolved, fundamentally important problems within the paradigm of modern physics with general relativity and quantum mechanics as its two main theoretical pillars.)

Having noticed the serious and irreparable inconsistency of general relativity and quantum mechanics, it seems neither unusual nor irrational that one can easily think of or think over: because the actual implication of such an inconsistency is that either or both of these two crucially important theories cannot be correct, are they really able to solve the two famous big problems—*dark matter* and *dark energy*? Or could they even be the shackles and obstacles to solving these two big problems? With this kind of rational and closely related questions as a clear clue or noticeable reminder, one could be psychologically prepared to accept the new theories (even they may be radically new) that are capable of solving the problems of *dark matter* and *dark energy*. Thus, *A Brief History of Time*, via its impartial witness to this serious and irreparable inconsistency, (though perhaps unexpectedly) can enable us, especially some of those respected and related professional people in physics and astronomy, to be less psychologically reluctant in accepting the revolutionarily new theories—as long as they really and truly have the ability to solve the fundamentally important problems of *dark matter* and *dark energy*, no matter who has developed these new theories; regardless of who has discovered these new theories. (Narrator: undoubtedly, this kind of possible role, when viewed from the large perspective of science development, is fundamentally or extraordinarily important, thus should be highly valued.)

(Reason 7) *A Brief History of Time (ABHOT)*, through the plain *reality* that it turns out to be reasonable and fair that one couldn't really understand *ABHOT*, has actually put a big question mark on the entire theory system of modern physics. Why? Today's physics, or modern physics, is facing *several dozens* of long-term unsolved, fundamentally important problems, such as dark matter, dark energy, gamma ray bursts, and ultra-high-energy cosmic rays, the GZK paradox (this famous paradox has been concisely explained in the Glossary of this book), the Pioneer anomaly (about this anomaly, please see the Glossary of this book if necessary), as well as high-temperature superconductivity, ... and so on. With this plain *reality* as a specific reminder, it seems neither unrealistic nor irrational that some insightful and intelligent experts in physics can naturally or easily think of and ask the rational and sensible question like: what if modern physics doesn't have the ability to solve these fundamentally important, also crucially important, problems in fact? With such a big question as a marked clue or as an easily perceptible hint, some, even many, insightful and sharp professional people in physics seem to have no difficulty thinking of and realizing: the *fact* that modern physics has been facing these long-term unsolved, fundamentally important problems can be quite comparable to the situation in which an enterprise has permanently insolvent debts in business. Clearly, such a realization actually tells or advises us that one shouldn't feel too surprised if today's physics really needs a great revolution. Thus, *ABHOT*, through this plain *reality*, seems to have played a positive role in knocking at the door of a revolutionarily new era of physics—at least in a sense or to a certain extent, even though such a positive role is apparently not the intended or original purpose of *ABHOT*. (Commentator: even only for this possible role, *ABHOT* shouldn't have any feelings of self-pity at all. Moreover, this possible role, when viewed from a large and historical perspective, can become rather rational or quite realistic, thus pretty feasible. Why? How? The answers to them are as follows.)

Historically, there were two clouds over the head of classical physics, which were the two famous problems that classical physics couldn't solve. One was that Newton's classical idea of absolute time and space was incapable of interpreting the observed phenomenon that the speed of light is the same for all observers, no matter whether they are moving or not with respect to the source of light. Another was that the classical theory of electromagnetic waves was unable to explain the energy radiation from a hot object or body, like an oven or a star, as just mentioned above. It was those two clouds that had caused or aroused the radical revolution from classical physics to modern physics. (Related information: modern physics, developed after 1900, with general relativity and quantum mechanics as its two main theoretical pillars, is ultimately based on several dozens of assumptions, hypotheses and postulates—AHPs, for short. For instance, general relativity hires at least five AHPs, and quantum mechanics employs at least four AHPs, as listed and pointed out in chapter one. Accordingly, to be objective and fair, the real or accurate name of modern physics is supposed to be postulate-based modern physics.)

By comparison, since only the two clouds over the head of classical physics could bring about the great and historic revolution from classical physics to postulate-based modern physics in the early 20th century, why can't today's *several dozens* of heavy and big clouds, being the several dozens of long-term unsolved, fundamentally important problems in physics and astronomy—as just mentioned, herald the epoch-making revolution from postulate-based modern physics to mechanism-revealed physics in the 21st century? (Commentator: with the insightful and highly positive comment like that, *A Brief History of Time* should not only throw out all its feelings of self-pity, but also be full of confidence.)

All in all, from a far-sighted perspective, *ABHOT* (the initials of *A Brief History of Time*) could have profound and far-reaching implications for the way in which we view the bright prospects for the further great developments of physics. *ABHOT*, through the "harsh" *reality* that it turns out to be rational and fair that one couldn't really understand *ABHOT*—many, even most, people have felt that it's actually impossible or extremely difficult to have a real understanding of *ABHOT*, though probably unintentionally, has signaled that physics seems to be at the eve of a great revolution. In this sense, *ABHOT* should feel gratified with itself! *ABHOT*, through its impartial *witness* to the serious and irreparable inconsistency of general relativity and quantum mechanics, though perhaps not on purpose, has forecast that physics seems to be approaching the dawn of a grand revolution. In this sense, *ABHOT* should be confident in itself! *ABHOT*, through the combined and mutual effect of such a *reality* and such a *witness*, though perhaps having not anticipated, has actually heralded that the splendid, epoch-making revolution in physics seems to be of a scientific and historic inevitability. In this sense, *ABHOT* should be content with itself! In other words, *ABHOT* seems to have reawakened us for being prepared to enter a grandly new era of physics—at least in a sense or to some extent. In this sense, *ABHOT* should be highly valued and fully respected!

(Commentator: YES! The grandly new era of physics is saluting to all of us—to men and women, to ordinary people and specialists, to teenagers and adults, and to all the people in the world. The future of this grandly new era is magnificently brilliant. This magnificent brilliance has great and radiant charm that is fascinatingly irresistible to all of us. So let us together wisely and enthusiastically embrace such a grandly new era!)

With these, and by what we have discussed above, before wrapping up this chapter and this book, I am delighted to reveal such a pleasant, plain truth: if you are the people who couldn't really understand *A Brief History of Time*, not only shouldn't you have felt discouraged or disappointed for that, but also ought to congratulate on yourselves for that! With the revealing of this truth, the people, especially the young people (the students in middle or high schools or colleges, for example), who had tried but failed to understand *A Brief History of Time*, can naturally and readily become the main driving force in the coming grandly new era of physics and its related other sciences. With their vigorous participation, the prospects for this grandly new era, by widely and greatly benefiting our entire humanity, can and will be even more magnificently brilliant. Let us together enjoy our wonderful lives under the marvelously magnificent brilliance of this grandly new era! The gloriously brilliant future of this grandly new era belongs to our entire humanity.

These wonderful and magnificent prospects also seem to remind us: we shouldn't forget the contribution of *A Brief History of Time*, even though this contribution is due to the reality that it turns out to be rather rational thus quite understandable that one couldn't really understand *it*. Someday, when these splendid and bright prospects become realistic, we ought to still remember the credit of *A Brief History of Time*, in spite of the fact that this credit is closely related to its witness to the serious and irreparable inconsistency of general relativity and quantum mechanics.

GLOSSARY

Acceleration: the rate at which the velocity of an object changes with time.

Astronomy: the branch of science that studies the sun, stars, planets, moon, etc.

Atom: the basic unit of matter that consists of a dense central nucleus surrounded by orbiting electrons.

Big bang singularity: being a product of general relativity, it is defined as a mathematical point that was infinitesimally small (practically zero size) with infinite density and infinite temperature.

Black hole*: a region of space having a gravitational field so strong that no matter or light can escape. (*Such a definition of black hole is the impassable obstacle to revealing the secrets of dark matter!)

Certainty mechanism behind the uncertainty principle: a newly revealed mechanism of science that shows that the *certainty* mechanism behind the *uncertainty* principle turns out to be the newly established and verified electron-photon work-energy relationship; this relationship is the most essential connection between an electron and its emitting photons. Over the range of one complete impulse period of the electron, the value of this certainty mechanism is equal to that of Planck constant h. (The uncertainty principle is about the electrons that emit photons, because this principle was developed based on quantum mechanics that describes how electrons radiate photons — electromagnetic radiation.)

Color index: the ratio of the wavelength of violet to red light.

Compton scattering: the experiment in which the high-energy rays of light, X-rays, were used to strike electrons.

Cosmic rays: also known as cosmic particles, made of electrons, protons, gamma rays and atomic nuclei, are the energetic particles originating outside of the earth. Cosmic rays travel at nearly the speed of light.

Cosmological arrow of time: the direction of time in which the universe is expanding, being "explained" by Stephen Hawking in his book *A Brief History of Time*.

GLOSSARY

Cosmological constant: an important constant originally in general relativity.

Cosmology: the branch of science that studies the universe, such as its origin, structure and development.

Crux: the most important or difficult part of a problem, a matter or an issue.

Dark matter: the huge amount of matter that exists in the universe, such as in galaxies and clusters. The amount of dark matter is estimated to be about 5.5 times of that of the ordinary matter (such as the matter of stars and planets) in the universe. The existence of dark matter cannot be observed directly but can be detected by its noticeable gravitational effects.

Electromagnetic radiation: the phenomenon of electrons emitting photons.

Electron: a stable elementary particle with negative electric charge that orbits the nucleus of an atom.

Electron-photon wavelength relationship: (a newly established and verified relationship of science) the ratio of the wavelength of an electron to the wavelength of its emitting photons is equal to the ratio of the orbiting velocity of the electron to the speed of light.

Electron-photon work-energy relationship: (a newly established and verified relationship of science) the work accepted by an electron over its one impulse period is equal to the energy of the photon emitted by the electron in the impulse period. This relationship reveals the most essential connection between an electron and its emitting photons.

Electron-self-exerting magnetic force: a newly discovered force, which is the magnetic force that acts on an orbiting electron through the magnetic field generated by the orbiting motion of the electron itself. The direction of this force is radially outward from the nucleus orbited by the electron to this electron.

Energy conservation (law): the law of science that states that energy cannot be created or destroyed, but only changed from one form into another or transferred from one object to another.

Event horizon: the boundary of a black hole.

Field: the area or space within which a specified force can be felt or has an effect; for example, the earth's gravitational field is the space in which the earth's gravity can be felt or has an effect.

Frequency: the number of waves per second when it is used to describe a wave.

Gamma rays: the highest energy and the shortest wavelength of electromagnetic radiation, which can be generated by nuclear reactions.

Gamma ray bursts (GRBs): the extraordinarily intense bursts or flashes of gamma rays in a very short time from an unknown source (or at least unknown within the paradigm of modern physics). These bursts, coming from different parts of the sky and occurring every day, can last from a fraction of a second to up

GLOSSARY

to a few minutes. The amount of energy released in a gamma ray burst is equivalent to all of the energy stored in the sun, so GRBs are known to be the most powerful explosions in the universe.

General relativity: Einstein's theory developed in the early 20th century, it describes and explains the force of gravity with the curvature of a four-dimensional space-time. This theory tells us that space and time are variable thus relative in a gravitational field, and that time runs slower in a gravitational field. General relativity is universally known to be seriously inconsistent with quantum mechanics.

Gravitational redshift: the redshift of light due to a gravitational field.

Gravitational scales of space and time: the space scale and time scale in a gravitational field.

Gravity: the force that attracts objects in space towards each other; the gravity of the earth makes things fall to the ground when they are dropped.

Impulse frequency of an electron: the number of complete impulses of an electron per second.

Impulse period of an electron: the time taken per complete impulse of an electron, which is the inverse of the impulse frequency of the same electron.

Magnetic field: a region around a magnetic material or a moving electric charge within which a charged, moving particle experiences magnetic force.

Magnetic force: a force that exists between two electrically charged moving particles, or the force exerted between magnetic poles.

Mass: the quantity of matter that an object contains; the mass of an object is measured by its acceleration under a given force or by the force exerted on it by a gravitational field.

Mass consumption: a basic principle or direct result that comes from the newly discovered law of mass doing work. This basic principle shows that the mass of an object decreases with the increase in its velocity, by revealing why and how the object's mass is being consumed due to its doing positive work.

Mass-energy equivalence equation: which is $E = mc^2$, where c is the speed of light, m is the rest mass of an object, and E is the rest energy of the object.

Mechanism-revealed black hole[**]**:** the black hole explained with the newly established mechanism-revealed black hole theory. ([**]Such a definition of black hole turns out to be the key to revealing the secrets of dark matter.)

Mechanism-revealed black hole theory[+]**:** a newly established theory of black hole that reveals the mechanism behind the observed phenomenon of black holes. According to this new theory, a black hole is a region where all visible light becomes invisible; in such a region, the existence of a hugely massive celestial (heavenly) body reduces the scales of length and time to such an extent that all visible light becomes invisible. So a black hole is simply a hugely massive celestial body; the mass of a black hole is greater or equal to several dozens of times

GLOSSARY

that of the sun. (⁺This new black hole theory is indispensable to solving the fundamentally important problems of dark matter.)

Mechanism-revealed gravitational theory: a newly established and verified theory of science that solves the fundamentally important problem of *why* and how space and time are variable thus relative in a gravitational field, by revealing and determining *why* and how their scales are variable thus relative.

Mechanism-revealed scales relativity theory: a newly established and verified theory of science that shows why time runs slower and why length becomes shorter in the situation of high speed, and that reveals the mechanism behind the two postulates of special relativity.

Mechanism-revealed theory: a theory of science that finds the mechanism behind its describing phenomena.

Momentum: a quantity of motion of a moving object, measured as its mass multiplied by its velocity.

Momentum conservation (law): the law of science that states that the total momentum of the objects of a system is constant if there are no external forces acting on the system. In such a system, the total momentum of two objects before a collision is equal to the total momentum of the two objects after the collision.

Nucleus: the positively charged central core of an atom, consisting of protons and neutrons, contains nearly all its mass.

Orbiting velocity of electron: the velocity along the direction of the position of the equilibrium orbit of the electron.

Photon: a tiny and discrete quantum of light (photons: the tiny and discrete quanta or particles of light).

Photoelectric effect: the phenomenon of electron ejection by light; when high-energy light shines on a metal surface, electrons are ejected from the metal surface.

Planck-Einstein equation: which is $E = hf$, where h is Planck constant, f is the frequency of a photon, and E is the energy of the photon.

Planck's quantum theory: the idea that light (or the energy released in electromagnetic radiation from electrons) can be emitted or absorbed only in discrete quanta—photons, whose energy is proportional to their frequency.

Prerequisite: an important thing required as a necessary or indispensable condition for something else to happen, exist, or be done.

Proportional: 'Y is proportional to X' means that Y is equal to X being multiplied by any constant number. For example, $Y = 3X$ is an expression of 'Y is proportional to X'.

Quantum: a discrete, indivisible quantity or unit of energy, especially the energy released in electromagnetic radiation from electrons.

GLOSSARY

Quantum mechanics: the theory developed based on the assumption that all forms of energy out of electrons are released in discrete units called quanta, which are known as photons nowadays. Since quantum mechanics is actually wave mechanics, quantum mechanics is, by using wave mechanics, to describe how electrons radiate photons.

Redshift of light: the measured wavelength of light becomes longer and longer.

Relativistic mass: a concept that comes from special relativity. This concept says that the mass of an object increases with the increase in its velocity, and the mass of an object becomes infinite large when the object infinitely approaches to the speed of light. So this concept is often simply stated as 'mass increase with speed'.

Space-time: the concept of one-dimensional time and three-dimensional space (being intermingled together) is regarded as a four-dimensional system.

Special relativity: a theory developed by Einstein at the beginning of the 20th century; it has two core concepts, time dilation and length contraction, respectively for interpreting time runs slower and length becomes shorter that appear in the situation of high speed. This theory initiates the era in which time and length are variable thus relative at different speeds.

Singularity: a mathematical point predicted by general relativity, whose size is zero with infinite density and infinite temperature. At this point, the curvature of space-time becomes infinite.

The essence of Planck constant: (a newly found essence of science or physics) the work accepted by an electron over its one complete impulse period is equal to the energy of the photon emitted by the electron in this impulse period.

The GZK paradox: one of the long-term unsolved fundamental puzzles in physics or astrophysics. The GZK paradox comes from GZK limit (computed by Greisen, Zatsepin and Kuzmin in the 1960s), which is a theoretical upper limit on the energy of cosmic rays from distance. GZK limit says that the distant cosmic rays with energies greater than certain threshold value should never be observed on the earth. However, a number of observations appear to indicate cosmic rays from distant sources with energies above this threshold value. So the key to solving the GZK paradox lies with finding out the sources of these cosmic rays in our Milky Way galaxy.

The law of an orbiting electron with periodic impulses emitting photons: a newly discovered law of science that solves the fundamentally important problem of *why* there are quantum states, by revealing the *quantum* mechanism of *why* and how photons, being the tiny and discrete quanta or particles of light, get their velocity c (the speed of light) from the electron emitting them.

The law of mass doing work: a newly discovered law of science. The core principle of this law is that the amount of energy in the mass of an object is measured by the amount of work done by the object's mass. This law shows that, when the velocity of an object is increased, the object's mass does positive work, the object thus loses the same amount of energy as that of the work done by the mass of the object from and by consuming its mass. As a result, an object's mass doing positive work causes a corresponding decrease in the object's mass.

The mechanism of Planck constant: (a newly revealed mechanism of science or physics) over one complete impulse period of an electron that emits photons, the amount of work accepted by the electron times its impulse period is equal to the value of Planck constant.

The mechanism of the famous mass-energy equivalence equation: the rest energy of an object, being the total energy contained in the rest mass of the object, is equal to the maximum ability of the object's mass doing positive work. (This famous equation is $E = mc^2$, where c is the speed of light, m is the rest mass of an object, and E is the rest energy of the object.)

The Pioneer anomaly: a small constant sunwards acceleration experienced by the Pioneer spacecraft unable to be explained with the laws of Newton and Einstein.

The postulate of 'equivalence principle': this postulate says that gravitational force has the same effect in increasing the velocity of an object as other traditional forces (this postulate, being a fundamentally indispensable postulate of general relativity as its fundamental and crucial foundation, is known to be the heart and soul of general relativity. In fact, this postulate is also known to be one of the most influential and famous postulates in modern physics).

Ultrahigh-energy cosmic rays: the cosmic rays that have extraordinarily high energy. Looking for the mysterious sources of ultrahigh-energy cosmic rays has been widely recognized by scientific community as one of the greatest unsolved fundamental puzzles in physics or astrophysics at present.

Uncertainty principle: a principle established by Heisenberg in the 1920s based on quantum mechanics. This principle says that one cannot exactly know both the position and the velocity of a particle (such as an electron) at the same time—the more accurately one measures the electron's position, the less accurately one can know its velocity, and vice versa.

Wavelength: the distance between two adjacent crests or two adjacent troughs.

Wave/particle dual natures of light: the concept says that the nature of light has two different aspects—wave aspect and particle aspect.

Weight: the force exerted on the mass of an object by a gravitational field. The weight of your body on the earth is the force exerted on the mass of your body by the gravitational field of the earth.

Wormhole: a theoretical thin tube of space-time that is assumed to connect distant regions of the universe (wormhole is also known as Einstein–Rosen bridge).

ACKNOWLEDGMENTS

My sincere acknowledgment goes to respected Dr. Stephen Hawking for clearly pointing out such a solid *fact* in his book *A Brief History of Time (ABHOT,* for short): general relativity and quantum mechanics cannot both be correct, because they are known to be inconsistent with each other (1998 version, P. 12). This solid fact is quite helpful to revealing the secrets of why it's really difficult to understand *ABHOT,* because these two famous theories virtually pervaded and penetrated everywhere in *that book.* This solid fact opens a bright window, through which dear readers can see the revealed secrets more clearly. This solid fact turns out to be one of the best ways for dear readers to know about these revealed secrets clearly, easily, and quickly, as well as enjoyably.

This solid fact actually tells us such a plain reality: either general relativity or quantum mechanics or both cannot be correct. This plain reality can substantially help dear readers grasp the crux or essence of the big puzzle of why it's really and truly difficult to understand *ABHOT,* because its most parts are, not only essentially but also crucially, based on general relativity and quantum mechanics. This plain reality can enable one to be well psychologically prepared on seeing the crux of why it's really not easy to understand *ABHOT.* This psychological preparation is quite positive for one to face and realize this crux calmly and rationally.

My sincere acknowledgment goes to distinguished and respected Dr. Hawking for explicitly pointing out that general relativity and quantum mechanics are known to be inconsistent with each other. (Friendly reminder: this inconsistency has profound consequences or crucial implications, because these two theories, general relativity and quantum mechanics, are the two main theoretical pillars of modern physics.) The direct reason behind this inconsistency turns out to be a clear clue pointing to the secrets of why it's really difficult to understand *ABHOT.* The deepest reason causing this inconsistency turns out to be one of the shortest routes leading us quickly to get to the crux of why it's really and truly hard to understand *ABHOT.* The exact reasons behind this inconsistency turn out to be a noticeable help for one to know about the secrets and grasp the crux of why it's really not easy to understand *ABHOT.*

ACKNOWLEDGMENTS

Because this book (*A Wonderful Gift to the Readers of "A Brief History of Time"*) is a popular science book, my top priority, also being one of my most important concerns, is to make it a highly readable book, so that most readers could really understand it thus truly enjoy it. In this aspect, my special acknowledgment goes to Bert Zhao and Briana Zhao, two students at Pullman High School, for having earnestly and attentively read this book before publishing it. Both of them have told me that they could easily understand almost all the materials in this book. Their comments have enabled me to have such a realistic estimate or reasonable confidence: most readers will find that they can clearly and easily understand what this book explains, clarifies and describes, or at least the overwhelming majority of dear readers will have no difficulty following the information presented in this book. In addition, the great efforts of these two high school students have played an important role in reducing the possible mistakes or errors (in language) of this book to a minimum level.

Bingcheng Zhao

INDEX

Absolute time 35, 270, 275
Acceleration 40, 60, 76, 184
Albrecht, Andreas 150
Anthropic principle 151, 222
  See also Weak anthropic principle
Arrow of time 217-225
Aristotle 121
Assumed quantum theory of gravity 145, 147, 149, 217-22, 252-53
  See also Quantum gravity theory; Quantum gravity
Assumptions, hypotheses and postulates (AHPs, for short) 4-7, 9-10, 125, 241-43, 245, 251, 275
Astronomy 78, 83, 93, 125, 129, 131, 271, 274-75
Astrophysics 83, 131, 271
Atom 46, 158, 167, 176, 190, 199
Atoms 11, 158, 257
Average density 131
  See also Density

Besso 46
Big bang 79-85, 134-35, 139-40, 143-45, 150-52, 223
Big bang singularity 79-82, 85
  See also Singularity
Bohr, Niels 46, 158, 161, 163, 167, 179, 190, 199
Bridges 228

Centripetal force 172
Certainty mechanism 153-56, 163-66, 180, 191-98, 201-02, 206-11, 214-16, 233-35, 246-49, 253, 273
Certainty mechanism behind the uncertainty principle 153-56, 163-66, 180, 191-98, 201-02, 206-11, 214-15, 233-35, 246-49, 253, 273
Certainty mechanism equation 192-93
Celestial bodies 83-84, 131, 139-40, 151, 223
  See also Heavenly bodies
Celestial body 74, 95-97, 116, 127-29, 131

Chaotic inflationary model 150
Classical physics 52, 54, 158, 175, 270-71, 275
Clouds 152, 270, 275
Colin 84
Color index 128
Compatible 3, 21, 28, 35, 38-42, 49, 58, 62, 145, 149, 176, 239, 257-58, 260, 266-67
Complain 255-56
Complaints 72, 256-57, 261, 263-68
Complementarity principle 161, 178
Compton, Arthur 120
Compton scattering 119-20, 167, 178, 260
Continent of North America 52, 59, 61, 76, 156, 233, 247, 249
Contour lines of space and time 95-96
Contradictory 2, 20-21, 23, 25-31, 46, 48-49, 58, 73, 80-81, 85, 101, 108, 113, 137, 171, 223, 231, 263, 272
Copenhagen 163
Core concepts 21, 30-31, 38, 73, 108, 258, 265, 267
Core constant 199
Cosmological arrow of time 217-22, 224-25
Cosmological constant 75
Cosmology 271
Coulomb force 157, 170
  See also Electric force
Crux 1, 10, 17, 25, 28, 60, 62, 65, 73, 75, 79-80, 82, 85-86, 92, 114, 120, 122-24, 130-32, 163, 207, 220, 240
Current method of calculating the momentum of photons 116-21
  See also Momentum; Momentum conservation law

Dark energy 102, 273-274
Dark matter 78, 88-89, 91, 102, 110, 123, 130, 141, 273-74
Davisson, Clinton 188
Davisson-Germer experiment 188, 199
de Broglie, Louis 158, 199
Deepest disaster 2, 148, 155, 164,

285

INDEX

215, 220
Density 74, 80, 85, 112-13, 131, 227-28, 232-33, 235
de Sitter, Willem 67
Diamond 61, 156
Dilemma 150, 221
Dimension 47
Dimensional 47
Dimensions 47, 57, 97
Dirac 160
Disappointed 1, 15, 18, 48, 103, 132, 143, 152, 221, 236, 255-56, 276
Discouraged 1, 15, 18, 48, 103, 132, 143, 152, 221, 255-56, 276
Dispel 263-65, 268
Disperse 263, 265, 267
Dissipate 256-57, 261, 263, 267-68
Doppler effect 65, 86-87, 89-92, 140-42, 144, 223-24
Duality 161, 179, 209, 249, 259, 262

Earth 7, 33, 36, 65, 71, 86-88, 91-92, 94-96, 141, 171, 177-78, 224, 239, 259
Einstein 46-47, 53, 118, 120, 162, 167-68, 175, 178, 180, 184-86, 189-90, 196, 199, 203-04, 228, 263, 270
Einstein, Albert 53, 162, 186
Electric field 170, 172
Electric force 157, 159, 170, 172, 175-76
Electromagnetic force 168
Electromagnetic radiation 157-59, 175-76
Electromagnetic waves 177-78, 270, 275
Electromagnetism 169, 171, 178
Electron or electrons 5-6, 8-9, 11, 45-47, 118-19, 146, 157-61, 166-76, 179-95, 199-206, 208-10, 212-14, 219, 233-34, 243-44, 247-48, 252, 257, 260, 270, 273
Electron-photon momentum relationship 118-19
Electron-photon wavelength relationship 187
Electron-photon work-energy relationship 180-81, 184-98, 200-06, 208-10, 233, 247-48
Electron-self-exerting magnetic force 169-72
See also Magnetic force

Elements 30, 157, 169, 227
Energy conservation law 3-4, 6, 185
Entropy 221
Equilateral triangle 184
Essence of Planck constant 195, 198, 201-06, 234, 248
Ether or "ether" 102, 177, 259
Event 18, 157, 177-79, 189, 207, 245, 259 (events 120)
Event horizon 111, 115-17, 119-23, 129-31
Event horizon of a black hole 115-16, 129, 131
Expanding universe 65-67, 70-71, 73, 75, 78-92, 98-99, 101-03, 133-35, 138-42, 144, 150-51, 218, 222-25
Experimental tests 5, 7, 9, 12, 31, 44, 53, 71, 136, 146, 212, 230, 243-44

Factory HHH 139
Feynman 162-63
Feynman, Richard 162
Force 5, 40, 75-76, 97, 101, 116, 157, 159, 168-72, 175-76, 182, 184, 251, 276
Forces 5-6, 75-76, 101, 169-70, 172, 176
Franco, Fernando 5
Frequency 74, 118, 126, 180, 184-85, 187, 189-91, 199-200, 203-04
Friedmann 67, 84-85
Friedmann, Alexander 67, 84

Galilei, Galileo 121
Galaxies 11, 86, 88-91, 139, 141, 151, 223, 257
Galaxy 11, 88-90, 131, 141, 151
Gamma ray bursts 78, 83-84, 123, 130-31, 139-40, 223, 274
Gamma rays 84, 131, 160
General knowledge or common sense 7-9, 29, 44, 51, 67, 71, 74, 93, 136, 138, 146, 155, 160, 168, 210, 212, 230, 243-44, 269
General relativity 1-7, 9-15, 17-20, 29-32, 48-49, 62-63, 65-82, 85-86, 92-94, 98-103, 105-12, 114-16, 121-23, 125, 130, 132-33, 135-39, 142, 144-46, 148-49, 151-56, 159, 164, 210, 214-15, 217, 219-23, 227-32, 235-36, 239-46, 250-51, 253, 256-61, 263-73, 275-76
General theory of relativity 79

INDEX

George 35-37, 127
Gravitational field 5-7, 9, 12-15, 18, 68-69, 71-72, 87-102, 107, 110, 114, 125, 127-29, 132, 135-36, 138, 141, 145, 148, 151-52, 219-20, 222, 224, 229-32, 242-43, 253, 264-67, 270-71
Gravitational force 5, 75-76, 97, 101, 251
  See also Force
Gravitational length scale reduction 267
Gravitational lens 89
Gravitational redshift 87-92, 141-42, 224
  See also Redshift; Redshift of light
Gravitational scale contour lines of space and time 95-96
Gravitational scales of space and time 95, 129
Gravitational time scale reduction 267
Gravitons 240, 251-53
Gravity 89, 95, 115-16, 145, 147-49, 217-22, 252-53
Great-great-great grandfather 227, 236
Guth, Alan 150
GZK paradox 123, 130, 274

Harvard tower experiment 71, 87, 94
Hawking, Stephen 2, 30, 79, 105, 107, 145, 218, 221
Heavenly bodies 83, 96, 151
Heisenberg 46, 157, 160-61, 167, 196, 199
Heisenberg, Werner 157, 160
  See also Uncertainty principle
Heterotic string 251
High-temperature superconductivity 274
Hot big bang model 134-35, 139-40, 143-45, 150-52
Hubble, Edwin 86, 142
Hubble's law 86, 142
Hydrogen atom 46, 158, 167, 175-76, 190, 199

Imaginary numbers 149
Imaginary time 149
Impulse frequency 187, 190-91, 200
  See also Frequency
Impulse period 119, 174, 180, 182-83, 185, 188-90, 192-95, 200-03, 233-34, 247
Incompatible 2, 3, 21, 23, 25-28, 31, 41, 48, 58, 73-74, 80-81, 85, 101, 108, 112, 137, 146, 148, 179, 223, 231, 257-63, 265, 272

Incompatibility 2, 28, 41, 148, 179, 240-41, 245-46, 250-51
Inconsistent 1-5, 10-11, 14-15, 17, 20, 29-31, 62, 66-67, 74, 79, 105-06, 109-10, 113, 133, 138, 145-46, 148-49, 153, 155, 164, 214, 217, 219-21, 223, 227, 229, 232, 236, 239, 257-65, 268-69
Inconsistency 1-4, 6, 9-14, 18, 20, 28, 62, 65-68, 107, 110, 136, 138, 145-46, 148, 154-56, 159, 164, 210-11, 215, 219-20, 229-30, 236, 239-42, 244, 250-51, 253, 256-61, 263-70, 273-76
Indispensable 4-5, 7, 10, 18-19, 31, 67, 69, 71, 76, 78, 84-85, 94, 101, 115, 126, 136, 144, 146, 158, 166, 180-81, 184-86, 188, 196, 199-200, 203, 205, 209, 212-13, 227-28, 230, 234, 236, 241-43, 245-47, 252
Indispensably 7, 9, 44, 47, 69, 81, 121, 132, 136, 146, 166, 213, 230, 243-44, 251, 253, 264
Infinite density 74, 80, 85, 112-13
  See also Density
Instantaneous maximum velocity 172-74
Instantaneous maximum velocity of (the) electron 172, 174
Interference 112, 177, 209, 249, 259
Intermingle 241, 246, 250
Intermingled 2, 217, 221, 246
Invisible light 126-27, 129
Israel, Werner 107

Jennifer 36
John 35-36
Jordan 160
Joule's law 186-87

Kerr, Roy 107

Linde, Andrei 150
Length becomes shorter 21, 25, 28, 39-45, 48-50, 53, 56-58, 73, 96-101, 108, 137, 223, 230, 263, 266-67, 270, 272
LeMaitre, Georges 67
Length contraction 21-23, 25-28, 31, 34-36, 38-42, 47-48, 58, 60, 62, 73, 101, 108, 137, 223, 230, 263-65, 272
Length scale reduction 55, 57-58, 266-67

INDEX

Light bending 7, 71, 89, 94
Light bending around the sun 7, 71, 94
Light waves 102, 177, 259

Magnetic field 169-71
Magnetic force 159, 169-72, 175-76
Major premise 143-45
Mansion 164, 215, 239, 268-69
Mark 35-36
Mass consumption 54-55, 76, 97, 173
Mass scale reduction 55
Mass-energy equivalence equation 52-54, 57, 77, 100, 184
Mass increase with speed 33-36, 38, 40-42, 62
    See also Relativistic mass
Maxwell, James Clerk 178
Measurement 48, 161
Mechanism and essence behind Planck constant 200, 203, 205
Mechanism and essence of Planck constant 195, 198, 201-06, 234, 248
Mechanism behind Planck constant 201
Mechanism of Planck constant 202-03
Mechanism-revealed 56-57, 96, 99-100, 125-32, 266-67, 271, 275
Mechanism-revealed black holes 125, 129, 131
    Their mechanism 128
    Their essence 128
Mechanism-revealed black hole theory 125-32
    Its mechanism 128
    Its essence 128
Mechanism-revealed gravitational theory 96, 100, 125, 127, 129, 132, 267
Mechanism-revealed physics 271, 275
Mechanism-revealed scales relativity theory 56-57, 99, 266
Mercury's orbit 7, 71, 94
Microwave radiation 65, 82-86, 92, 139-40, 142, 144, 223
Milky Way galaxy 88-90, 131, 141
Model of the hydrogen atom 46, 158, 167, 175-76, 190, 199
Modern physics 2-3, 46, 76, 148, 155-56, 158-60, 162, 164, 167, 175, 210, 215, 220, 239-40, 245, 261, 268-71, 273-75
Murder case 14, 208, 233, 249
Momentum 5-6, 116-21, 161, 181-85, 187-88, 192-93, 195, 235, 245-46, 248
Momentum conservation law 6, 118

Negative energy density 227-28, 232-33, 235
Newton, Isaac 178
Newtonian gravitation theory 116
Newton's second law 5, 116, 184
New inflationary model 150
No boundary condition 218
No boundary proposal 217-22, 225
Nobel Prize 167, 188

Observational interference 209, 249
    See also Interference
Observational tests 5, 7, 31, 42, 71
Old inflationary model 150
Oppenheimer, Robert 107
Orbital frequency 191
    See also Frequency
Owner 239, 250-51, 253-54

Penrose, Roger 107
Penzias, Arno 83
Photoelectric effect 118, 120, 167-68, 175, 178, 199, 260
Photons 5, 7-9, 45-47, 74, 116-21, 130-31, 145-46, 157-59, 161, 166-68, 171, 173-76, 178-93, 195, 199-206, 208-10, 212-14, 219, 234, 243-44, 247-48, 252, 270, 273
Planck constant 118, 180, 184-85, 189-90, 192-95, 197-206, 208, 234-35, 248
Planck constant equation 200-05
Planck-Einstein equation 180, 184-86, 189-90, 199, 203-04
Planck, Max 158, 162, 186, 204
Planck's quantum theory 167, 178, 270
Postulate-based modern physics 275
    See also Modern physics
Pound-Rebka-Snider experiment 87-88
Prerequisite 4, 11, 40, 61, 68-69, 83-84, 86-92, 131, 134-35, 139-42, 145, 147, 181, 219, 223-24, 236, 241-42, 245-46, 250-53
Princeton University 87
Proportional 5, 116, 129, 186, 190
Proportionally 202-03
Prospects 227-28, 235-36, 239, 245, 275-76
Psychological hesitation 78, 110, 138, 144, 261, 269

INDEX

Psychological obstacle(s) 75, 110, 124, 229, 248
Psychological preparation 256
Psychologically comfortable 28, 32
Psychologically happy or ready 70, 229
Psychologically hesitant 71, 144, 269
Psychologically reluctant 29, 43, 223, 231, 236, 274
Psychologically uncomfortable 14

Quantum gravity 145, 218, 252
Quantum gravity theory 145, 218, 252
Quantum hypothesis 158, 162, 199
Quantum mechanics 1-6, 8-18, 20, 29-31, 46, 62, 65-68, 74, 79, 105-07, 109-10, 133, 136, 138, 145-49, 153-64, 166-67, 175-76, 196, 199, 208, 210-17, 219-21, 223, 227, 229-30, 232-34, 236, 239-45, 247, 250-53, 256-70, 272-73, 275-76
Quantum physics 161-63, 168, 188, 197, 199, 204, 210

Radius of a black hole 128-29
Railway tunnel 239
Random effects 208-09, 233, 246, 249
Random factors 209
Rate of expansion 144
Ratio of visible light boundary wavelength 128
Redshift 86-92, 140-42, 224
Redshift of light 87, 91, 141, 224
Region of space-time 227-28, 232
Reveal (+ revealed, revealing, reveals) 6, 8, 10, 14-15, 26, 32-33, 35, 41-43, 47, 50, 52-54, 56-62, 65, 69, 71, 73-79, 82, 84-85, 92, 94-96, 98-102, 107, 109-10, 114, 119, 122-25, 127, 129-33, 135, 137-40, 143-45, 147, 149-50, 152, 154, 156, 163, 165-66, 168, 173-78, 181-82, 184-85, 187-88, 190-98, 200-11, 214-16, 219, 222-25, 229, 231, 233-35, 237, 242-45, 247-49, 252-53, 255-57, 262-63, 266-67, 270-73, 276
Relationship 4, 5, 10, 56, 74-75, 118-19, 180-81, 183-206, 208-10, 233-34, 241-42, 246-48, 251
Relationships 40, 60, 182
Relativistic mass 33-36, 38, 40-42, 62
Root cause(s) 1, 11, 60, 62, 65, 68, 73, 96, 99, 101, 107, 114, 120, 132-133,
136, 138, 148, 209-211, 214-216, 219, 230, 240, 242, 257, 269
Rosen, Nathan 228
Rotation of the long axis of Mercury's orbit 7, 71, 94

Schrodinger 46, 157, 160, 162, 167, 196-97, 199
Schrodinger, Erwin 157, 162
Schwarzschild, Karl 106
Schwarzschild radius 129
Science Magazine 2, 147, 220, 252
Second law of thermodynamics 217, 221-22
Secrets 1, 50, 65, 92, 96, 98-99, 101-02, 123, 132-33, 152, 190, 237, 245, 255-56
Self-contradiction 25, 27, 38
Self-contradictory 17-21, 23, 25-35, 38-42, 46-50, 57-58, 60, 62, 73-75, 80-81, 85, 101, 108-09, 111-15, 121-23, 130, 137, 223, 230-31, 272
Self-evident 8, 19, 25-26, 30, 32, 39, 59, 61, 66, 68, 70-73, 75-76, 79-80, 94, 106, 108-09, 111-12, 123-24, 130, 137-38, 150, 168, 209, 215-16, 218, 221, 231, 246, 249, 259, 265
Singularity 74, 79-82, 85, 111-15, 122-24, 130, 145, 218
Size of a black hole 120, 124, 128-29
Sound waves 177-79, 259
Space-time 3, 5, 111, 227-28, 232
   See also Region of space-time
Special relativity 17-36, 38-50, 57-63, 73, 100-01, 108, 121-22, 137, 222, 230-31, 258, 263-66, 270, 272
Speed of light 8, 21, 27, 34-36, 40, 44-47, 52, 57-60, 74, 77, 80, 100, 112, 116, 118, 121, 129, 146, 166-68, 172-75, 184, 187, 191, 199, 210, 213-14, 219, 243, 252-53, 270, 273, 275
Stage of particle nature 178
Stage of wave nature 178
Stage of wave/particle dual natures 178
Star 87-88, 91, 114, 141, 158, 167, 178, 199, 224, 260, 270, 275
Stars 11, 65, 83, 86, 88-92, 131, 141, 151, 224, 257, 271
Statistical interpretation 162, 209, 249
Steinhardt, Paul 150
String theories 240, 250

289

INDEX

String theory 244, 250-51, 253
Sun 7, 71, 87-91, 94-95, 114, 131, 141, 177-178, 259
Supergravity theories 240, 244, 250-51, 253
Supermassive black hole 90
Superstring theory 250

Telescope 102
Temperature 40, 60, 74, 80, 85, 112, 195
'The Blind Men and the Elephant' 36-38, 87, 120
The constancy of velocity scale 58-59
The constant scale ratio of mass, length and time 60-61
The encyclopedia of physics 2, 147, 220
The law of an orbiting electron with periodic impulses emitting photons 118, 173-74, 176, 179-81, 185, 210, 214
The law of mass doing work 53-54, 57, 77, 100
The mechanism behind the first postulate of special relativity 58-59
The mechanism behind the second postulate of special relativity 60-61
The essence of Planck constant 201-203
The mechanism of Planck constant 202-03
The mechanism of the famous mass-energy equivalence equation 53-54
The Pioneer anomaly 78, 110, 274
The postulate of 'equivalence principle' 5, 75, 77, 101 (The newly revealed mechanism behind this postulate 76)
Theoretical pillars 2-3, 78, 110, 155, 161, 164, 215, 220, 239, 261, 268-69, 271, 273, 275
Theory of general relativity 63, 80, 82, 99, 106, 114, 137, 144, 228
Thermodynamic arrow of time 217, 221-22
The uncertainty in energy and time 195, 235, 248
The uncertainty in momentum and position 195, 235, 248
The World's 20 Greatest Unsolved Problems (a book) 2
Three fundamental constants 199
Threshold radius 128
Time dilation 21-23, 25-28, 31, 33-36, 39-42, 47-48, 58, 60, 62, 73, 101, 108, 137, 223, 230-31, 263-65, 272
Time runs slower 6, 12, 14, 21, 25, 28, 40-45, 48-50, 53, 56-58, 69, 73, 95-101, 108, 110, 135, 137, 223, 229-30, 242, 263-64, 266-67, 270, 272
Time scale reduction 55, 57-58, 266-67
Time travel 227-28, 231-33, 235-37
Twin Paradox 33-39, 42, 62

Ultrahigh-energy cosmic rays 83-84, 123, 130, 139-40, 274
Uncertainties 161-162, 195-198, 201, 206, 208, 234-235, 248
Uncertainty principle 153-57, 160-66, 180, 191-99, 201-02, 206-11, 214-16, 227, 232-36, 240-41, 245-51, 253, 257, 259-60, 262, 273
Unification of physics 239, 245, 261
Unification of wave/particle dual natures 179

Vector 182
Vector adding up 182
Vector superposition 182
Velocity 5, 8, 21, 27, 33-34, 40, 45-47, 52, 54, 58-60, 75-76, 101, 116-17, 121, 146, 161, 166-69, 172-75, 181-85, 187-89, 191-93, 202-03, 210, 213-14, 219, 243, 252, 270, 273
Velocities 182-83
Visible light 126-28

Water waves 178
Wavelength 118, 126-28, 158, 187, 192, 199
Wave mechanics 8, 146, 157, 159-60, 211-12, 243
Wave/particle duality 161, 179, 209, 249, 259, 262
  See also Duality
Wave/particle dual natures of light 173, 176-77, 179-80, 260, 262
Weak anthropic principle 222
Wilson, Robert 83
Work-energy relationship 180-81, 184-98, 200-06, 208-10, 233-34, 247-48
Wormholes 227-28, 231-33, 235-37

Young, Thomas 178

ABOUT THE AUTHOR

Bingcheng Zhao, who was born in 1963 in Shandong Province of China, obtained his Ph.D. in 2001 from Washington State University, Pullman, Washington State, the United States of America. He is also the author of *From Postulate-Based Modern Physics to Mechanism-Revealed Physics*, published in 2009. The newly founded and verified mechanism-revealed physics is the key to solving the fundamentally important problems of dark matter and dark energy.

www.ingramcontent.com/pod-product-compliance
Lightning Source LLC
Chambersburg PA
CBHW020855180526
45163CB00007B/2509